Agricultural Irrigation

Agricultural Irrigation

Special Issue Editor

Aliasghar Montazar

MDPI • Basel • Beijing • Wuhan • Barcelona • Belgrade

MDPI

Special Issue Editor
Aliasghar Montazar
University of California Cooperative Extension
USA

Editorial Office
MDPI
St. Alban-Anlage 66
4052 Basel, Switzerland

This is a reprint of articles from the Special Issue published online in the open access journal *Agriculture* (ISSN 2077-0472) from 2018 to 2019 (available at: https://www.mdpi.com/journal/agriculture/special_issues/Agricultural_Irrigation).

For citation purposes, cite each article independently as indicated on the article page online and as indicated below:

LastName, A.A.; LastName, B.B.; LastName, C.C. Article Title. *Journal Name* **Year**, *Article Number*, Page Range.

ISBN 978-3-03921-922-3 (Pbk)
ISBN 978-3-03921-923-0 (PDF)

Cover image courtesy of Aliasghar Montazar.

Contents

About the Special Issue Editor

Aliasghar Montazar holds a PhD in irrigation and drainage engineering, and is currently Irrigation and Water Management Advisor at the University of California Division of Agriculture and Natural Resources. He is a previous Project Scientist at the University of California, Davis, and a former Associate Professor at the Department of Irrigation and Drainage Engineering, University of Tehran. His research, teaching, and outreach activities focus on best irrigation and nutrient management practices, deficit irrigation strategy, viability assessment of on-farm water conservation practices and technologies, sensor-based agricultural water management and crop water use measurements, and the development of crop coefficients in California low desert cropping systems.

Preface to "Agricultural Irrigation"

Agriculture is certainly the most important food supplier, globally accounting for more than 70% of used water and contributing significantly to water pollution. Irrigated agriculture is facing rising competition worldwide regarding access to reliable, low-cost, and high-quality water resources. However, irrigation, as the major tool and determinant of affecting agricultural productivity and environmental resources, plays a critical role in food security and environment sustainability. Innovative irrigation technologies and practices may enhance agricultural water efficiency and production, simultaneously mitigating water demand and quality issues.

This Special Issue brings together a collection of ten cutting-edge studies focusing on recent advancements in agricultural irrigation which assess the current challenges and offer approaches, tools, and opportunities for the improvement of future irrigation. Haghverdi and co-workers report cotton crop yield response to irrigation and the spatial heterogeneity of soil physical attributes in a humid region. Zikalala and co-workers present a root zone salinity model for vegetable crops. My group and I report a preliminary viability assessment of using drip irrigation for organic spinach production and the management of spinach downy mildew disease. Francaviglia and co-worker outline the effects of deficit irrigation strategies on crop yields and irrigation water utilization efficiency of processing tomato. Handa and co-workers report the results of their investigation on the efficiencies, environmental footprint, and economics of irrigation pumping in two Oklahoma areas that rely heavily on groundwater resources. Bekmirzaev and co-workers outline the effect of irrigation water regimes on *Tetragonia tetragonioides* productivity in a semi-arid region. Lu and co-workers present a model developed for optimal sprinkler irrigation management in tea fields of the Yangtze River region. Muema and co-workers report a combination of benchmarking and principal component analysis for evaluating the efficiency of irrigation schemes in Kenya. Dalias and co-workers outline the required adjustment of irrigation schedules as a strategy to mitigate climate change impacts on agriculture in Cyprus. And, finally, Zheng and co-workers discuss an evaluation of management practices (cover crops, mulcting, and using city water for irrigation) on improving thermal protection with low input sustainable practices in lettuce production. I hope this Special Issue will be useful to researchers, professionals, and students, as well as contribute to developments in future irrigation.

Aliasghar Montazar
Special Issue Editor

agriculture

MDPI

Article

Studying Crop Yield Response to Supplemental Irrigation and the Spatial Heterogeneity of Soil Physical Attributes in a Humid Region

Amir Haghverdi [1,*], Brian Leib [2], Robert Washington-Allen [3], Wesley C. Wright [2], Somayeh Ghodsi [1], Timothy Grant [2], Muzi Zheng [2] and Phue Vanchiasong [2]

[1] Department of Environmental Sciences, University of California, Riverside, 900 University Avenue, Riverside, CA 92521, USA; somayehg@ucr.edu

[2] Department of Biosystems Engineering & Soil Science, University of Tennessee, 2506 E.J. Chapman Drive, Knoxville, TN 37996-4531, USA; bleib@utk.edu (B.L.); wright1@utk.edu (W.C.W.); tgrant7@vols.utk.edu (T.G.); mzheng3@vols.utk.edu (M.Z.); vanchias@gmail.com (P.V.)

[3] Department of Agriculture, Nutrition, and Veterinary Science (ANVS), University of Nevada, Reno, Mail Stop 202, Reno, NV 89557, USA; rwashingtonallen@unr.edu

* Correspondence: amirh@ucr.edu

Received: 21 January 2019; Accepted: 20 February 2019; Published: 23 February 2019

Abstract: West Tennessee's supplemental irrigation management at a field level is profoundly affected by the spatial heterogeneity of soil moisture and the temporal variability of weather. The introduction of precision farming techniques has enabled farmers to collect site-specific data that provide valuable quantitative information for effective irrigation management. Consequently, a two-year on-farm irrigation experiment in a 73 ha cotton field in west Tennessee was conducted and a variety of farming data were collected to understand the relationship between crop yields, the spatial heterogeneity of soil water content, and supplemental irrigation management. The soil water content showed higher correlations with soil textural information including sand ($r = -0.9$), silt ($r = 0.85$), and clay ($r = 0.83$) than with soil bulk density ($r = -0.27$). Spatial statistical analysis of the collected soil samples (i.e., 400 samples: 100 locations at four depths from 0–1 m) showed that soil texture and soil water content had clustered patterns within different depths, but BD mostly had random patterns. ECa maps tended to follow the same general spatial patterns as those for soil texture and water content. Overall, supplemental irrigation improved the cotton lint yield in comparison to rainfed throughout the two-year irrigation study, while the yield response to supplemental irrigation differed across the soil types. The yield increase due to irrigation was more pronounced for coarse-textured soils, while a yield reduction was observed when higher irrigation water was applied to fine-textured soils. In addition, in-season rainfall patterns had a profound impact on yield and crop response to supplemental irrigation regimes. The spatial analysis of the multiyear yield data revealed a substantial similarity between yield and plant-available water patterns. Consequently, variable rate irrigation guided with farming data seems to be the ideal management strategy to address field level spatial variability in plant-available water, as well as temporal variability in in-season rainfall patterns.

Keywords: farming data; precision agriculture; site-specific irrigation

1. Introduction

1.1. Supplemental Irrigation Management in Humid Regions

Irrigated agriculture has been playing a globally significant role in providing roughly one-third of the total food and fiber supply [1]. While irrigated acreage is shrinking in some arid regions in the

US due to increasing competition for water, supplemental irrigation is expanding in humid regions as a means to avoid unpredicted periods of water stress and maintain high yields [2]. For example, in west Tennessee, row crop irrigation has expanded rapidly from twenty-five center pivot irrigation systems installed in 2007 to 270 systems installed in 2012. This represents an expansion of 16,000 ha of cropland per year under supplemental irrigation [3], which necessitates an essential demand to study supplemental irrigation management of different crops in this region.

Precipitation is the main source of moisture in west Tennessee. However, severe in-season drought conditions for short periods are likely to occur, which could substantially reduce yields under rainfed agricultural practices. Supplemental irrigation is an irrigation strategy that attempts to maintain maximum yield production by irrigating during periods of insufficient rainfall to fulfill the crop water requirements. The application of supplemental irrigation management is a complex problem in west Tennessee, where precipitation patterns are temporally variable within and across cropping seasons and interact with the spatial mosaic of the physical and hydrological attributes of alluvial and windblown loess deposited soils. Soil properties, such as texture and bulk density, greatly affect soil water retention and movement and govern readily available soil water for crop irrigation management. Excess water content within the root zone could occur if irrigation adds to unpredicted rainfall events. This may cause insufficient aeration and consequent yield reduction. Moreover, runoff and deep percolation may lead to accelerated nutrient loss and soil erosion that in turn, increases the risk of contamination of nearby surface and/or groundwater.

Crop yield has been proven to be strongly related to soil physical properties. For example, Ref. [4] considered plant-available water (PAW: volumetric water content between the field capacity and the permanent wilting point within the root zone) as an input predictor of the wheat yield. They reported PAW as one of the dominant factors governing the spatiotemporal variation of yields. Soil texture was discussed by [5] as one of the greatest factors affecting the cotton yield. They found a relatively stable spatial pattern of yield over time, although yield and soil properties had stronger relationships during dry seasons than wet seasons. Graveel et al. [6] studied the response of corn to variations in soil erosion and sandy and silt textured profiles in west Tennessee and found a substantial difference in yield.

Cotton is a major crop in west Tennessee that is grown in more than 15 states and is vital to the US economy because it is a critical export-oriented product [7]. Currently, some 40% of US cotton is under irrigation, with the area expanding throughout the mid to southern US. Given the limited water resources in many cotton-growing areas, a considerable amount of research has recently been performed on cotton irrigation to improve the water use efficiency [8]. However, inconsistent cotton yields have been observed in response to irrigation in the humid portion of the US [9]. Suleiman et al. [10] studied the use of cotton deficit irrigation in a humid climate using FAO's 56-crop coefficient method in Georgia and suggested establishing a 90 % irrigation threshold for the full irrigation of cotton in humid climates. Bajwa and Vories [11] evaluated the cotton canopy response to irrigation in a moderately humid area in Arkansas and found that under wet conditions, excessive irrigation decreased the yield of cotton lint. A similar result was reported by [12], who also found that excessive rainfall limited the yields from irrigation. Gwathmey et al. [13] conducted a four-year supplemental irrigation study in Jackson, Tennessee, and found a 38% improvement in lint yields at a 2.54 cm wk^{-1} supplemental irrigation rate compared to three of four years of the rainfed irrigation scenario. Grant et al. [14] used a surface drip irrigation system to investigate the response of the cotton yield to irrigation across different soil types with different PAW. This study illustrated that uniform irrigation is not the optimum management decision for the cotton wherever field-level soil heterogeneity affects the spatial distribution of PAW.

1.2. Farming Data and Precision Agriculture

Traditionally, irrigation studies were limited to small plots at research stations, mostly due to economic and computational limitations. Additionally, contemporary constraints to irrigation studies include the personnel time and expense for data collection, as well as the limitations of conventional computing infrastructure and statistical methods to analyze the increasingly larger spatiotemporal datasets that have inherent noise and uncertainty. In west Tennessee, the inherent heterogeneity and the spatiotemporal changes in soil and weather-related attributes of the region make it hard to extrapolate the results of design-based experiments on small plots to real field conditions. Supplemental irrigation scheduling is a site-specific irrigation management question where each field has its own irrigation management challenge that requires unique solutions. On-farm experimentation is an alternative for design-based experiments, since collecting site-specific information is becoming more and more common and affordable in US agriculture.

In contemporary agriculture, precision farming enables farmers to locally collect various site-specific information, such as the yield and soil apparent electrical conductivity (ECa). Crop yield maps provide valuable quantitative information on crop production, change in production, and the response of crop production to different agricultural inputs, including irrigation and fertilizer. Soil survey maps; soil sampling; on-the-go sensors; and remote sensing from field, airborne, and satellite sensors are the most widely used methods to obtain information on the spatial distribution of different soil attributes [15]. Soil sampling at the field-level provides valuable information on the spatial variation of soil attributes, but collecting this data has become laborious and expensive. ECa is a proxy for less accessible soil attributes, including soil texture and soil available water [16], and thus has created substantial interest in its use for soil mapping and management zone delineation in precision agriculture. ECa is measured in a simple and inexpensive way, where an electrical current is induced into the soil while the field is traversed. However, there are some inconsistencies in the literature concerning factors that affect the variability of ECa in non-saline fields [17]. This suggests the need to investigate the practical utility of using ECa for site-specific management in different regions, particularly because most of the supporting ancillary datasets including topographic edaphic features (e.g., elevation, slope, and aspect) are freely available. If not, these site-specific attributes can be measured and mapped without spending a considerable amount of time and money. Recently, new wireless technologies have enabled progressive farmers to remotely and continuously monitor soil properties over time, including soil temperature, soil water content, and soil matric potential.

Consequently, this study was carried out to understand the relationship between the spatial heterogeneity of soil and crop yields to better inform the management of site-specific supplemental irrigation in west Tennessee. The objectives of this study were to conduct an on-farm experiment and analyze yield maps to:

1. Assess the impact of the spatial heterogeneity of soil water content on the pattern of yield using on-farm data that was collected by the farmer's soil moisture sensors and yield monitor systems;
2. Compare the cotton lint yield under different supplemental irrigation regimes across different soil types;
3. Assess the temporal stability of low/high yield zones by combining the measured historical yield data of different crops with available cotton yield data.

2. Materials and Methods

2.1. Study Area

The study area was a 73 ha irrigated field that is located in southwestern Dyer County in west Tennessee along the Mississippi river (Figure 1). The field was equipped with two center pivot irrigation systems that were used for the irrigation of no-till cotton during each cropping season. The field is on Mississippi river terrace alluvial deposits from which Robinsonville loam and fine sandy loam, Commerce silty clay loam, and Crevasse sandy loam soils have been produced (Figure 1). Figure 2 illustrates the long-term variability in regional climate. The mean monthly growing season precipitation and temperature is 97-mm month^{-1} and 21 °C from May to November, respectively (Figure 2). Rainfall is relatively high, even in dry years. Temperature changes are less pronounced and to some extent, inversely proportional to rainfall. The supplemental irrigation strategy has been growing in this region since rainfall events are not usually temporally well-scattered to fulfill the crop water requirement over the entire growing season.

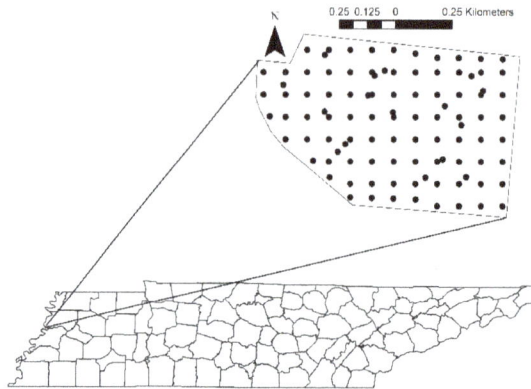

Figure 1. The 73-ha supplemental irrigation study field is located in southwestern Dyer County in west Tennessee along the Mississippi river. Soil samples were collected at four depths from 0–1 meter at 100 locations.

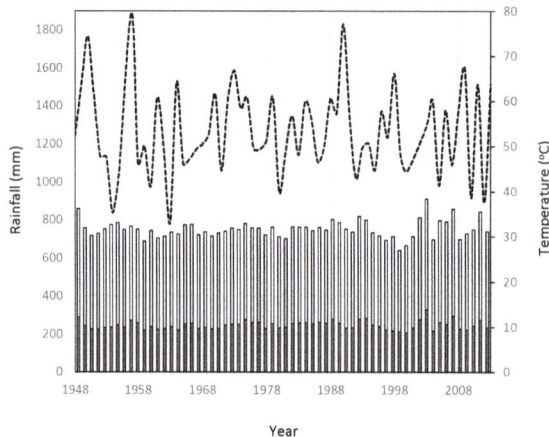

Figure 2. The long-term climatic variation in rainfall (dashed line) and temperature (column) in west Tennessee. Temperature columns show the mean monthly minimum (in black) and the mean monthly maximum (in white).

2.2. Soil Data Collection and Lab Analysis

Haghverdi et al. [18] described the soil data collection where one hundred undisturbed samples (100 cm deep) were collected by a truck-mounted soil sampler between 21 and 22 March 2014 (Figure 1). Some 86 of these samples were collected using a grid sampling scheme where samples were about 100-m apart (i.e., half the mean semivariogram range of proxies). The rest of the samples (=14) were randomly collected from underneath the center pivot circles. The field sampling occurred after rainfall events, when the soil water status was assumed to be close to the field capacity.

Each 67-mm diameter core was sub-sampled at four depths between 0–100 cm in 25-cm increments, i.e., 0–25 cm, 25–50 cm, 50–75 cm, and 75–100 cm, with adjustments in respect to the available horizons. The mean depth across samples approximated 25 cm for all the layers. Hereafter, the word "layer" is used to describe subsamples rather than real soil horizons. The soil texture of each depth was estimated in the laboratory using a hydrometer [19]. The soil water content was estimated by subtracting oven-dried weights from wet weights. Bulk density (BD) was estimated as the oven dry weight to volume of each subsample. ECa was collected using a Veris 3100 (Veris Technologies, Salina, KS, USA) instrument on March 20, 2014 with 10 m and 20 m spacing between points in the same row and adjacent rows, respectively. The Veris 3100 has six rolling coulters for electrodes and collects two simultaneous ECa measurements from shallow (~0–30 cm) and deep depths (~0–90 cm).

2.3. Descriptive and Spatial Analysis of Soil Properties

The correlation between the volumetric water content at the time of sampling and soil texture, i.e., sand, silt, and clay percentages, and bulk density was investigated. A soil texture triangle was plotted for each of the four depths, with each depth layer being approximately 25-cm thick. The relationship between ECa data and soil physical information, obtained from soil samples, was studied. To match ECa and soil basic data, the ECa data were interpolated to each sample using an ordinary kriging approach [20].

The spatial analysis was done using ARCGIS 10.2.2 [21]. To examine the spatial autocorrelation of the attributes, the semivariogram (Equation (1)) and Global Moran's I statistic (Equation (2), [22]) were obtained as follows:

$$\gamma(\mathbf{h}) = \frac{1}{2N(\mathbf{h})} \left\{ \sum_{i=1}^{N(\mathbf{h})} [Z(\mathbf{x}_i + \mathbf{h}) - Z(\mathbf{x}_i)]^2 \right\} \tag{1}$$

where $\gamma(\mathbf{h})$ is the semivariance; $\mathbf{h}$ is the interval class; $N(\mathbf{h})$ is the number of pairs separated by the lag distance; and $Z(\mathbf{x}_i)$ and $Z(\mathbf{x}_i + \mathbf{h})$ are measured attributes at spatial location i and $i + \mathbf{h}$, respectively. The nugget effect, sill, and range are the basic parameters of a semivariogram to describe the spatial structure. The nugget effect mostly represents sampling/measurement errors and variation at scales smaller than the sampling interval. The total variance is called the sill and the range is the maximum distance at which variables are spatially dependent.

The Global Moran's I statistic is calculated as:

$$I = \frac{n}{\sum_{i=1}^{n}\sum_{j=1}^{n} w_{i,j}} \times \frac{\sum_{i=1}^{n}\sum_{j=1}^{n} w_{i,j} z_i z_j}{\sum_{i=1}^{n} z_i^2} \tag{2}$$

where z is the deviation of an attribute from its mean, $w_{i,j}$ is the spatial weight between the ith and jth point, and n is equal to the number of points. Moran's I is used to measure the degree of spatial autocorrelation or trend based on both feature locations and feature values simultaneously. Given a set of features and an associated attribute, it evaluates whether the pattern expressed is clustered, dispersed, or random [22]. The null hypothesis of this analysis states that the attribute being analyzed is randomly distributed among the features in the study area. Ordinary kriging was applied to samples of the ECa to generate maps that were compared and assessed against each other. A higher positive Moran's Index for an attribute indicates a stronger spatial structure. The z-score changes in line with the Moran's Index. A z-score from −1.65 to 1.65 shows that the spatial pattern is not significantly different than a random one. A z-score less than −1.65 is an indicator of a dispersed process, while a z-score greater than 1.65 displays a spatially clustered attribute.

2.4. On-Farm Irrigation Experiment

There were two center pivot systems available for irrigation within the 73-ha field. The on-farm experiment was conducted for two years and designed to study the supplemental irrigation-cotton lint yield relationship across different soil types. The farmer used a no-tillage method to plant 'PHY375' cotton variety on 30 May 2013 and 'Stoneville 4946' on 5 May 2014. The farmer used soil test recommendations for applications of variable rate potassium (K) and phosphorus (P). However, nitrogen was applied uniformly. Crop pest management was implemented following state extension recommendations and the field was harvested on 2 and 3 December 2013 and in the second year on 18–20 October 2014.

Throughout the experiment, we used the Management of Irrigation Systems in Tennessee (MOIST) program (http://www.utcrops.com/irrigation/irr_mgmt_moist_intro.htm) to discuss the efficiency of irrigation management with the farmer. MOIST is an irrigation decision support tool that delivers irrigation recommendations by simultaneously measuring and monitoring soil water status and calculating water balance through a deployed wireless soil sensor network. An on-farm weather and soil monitoring station contained a number of METER Devices (METER Group, Inc., Pullman, WA, USA), including an EM50G remote data logger, a VP-3 temperature and relative humidity sensor, an ECRN-100 high-resolution rain gauge, and a pyranometer: a solar radiation sensor, was installed in 2013 and run through 2014 using the MOIST program. Three additional stations with rain gauges and soil moisture sensors were added in 2014. Each station also had two MPS-2 soil matric potential and temperature sensors (METER Group, Inc., Pullman, WA, USA) installed at approximately 10 and 46 cm depths to monitor the soil water status. MOIST calculates the daily reference evapotranspiration (ET_{ref}) using Turc's 1961 equation (developed for regions with relative humidity > 50%, [23]) as follows [24]:

$$ET_{ref} = 0.013 \times \left(\frac{T}{T + 15} \right) \times (R_s + 50) \tag{3}$$

where ET_{ref} is the daily reference evapotranspiration (mm d^{-1}), R_s is the daily solar radiation (Cal cm^{-2} d^{-1}), and T is the daily mean air temperature (°C). The data for each station were recorded once per hour, stored in the logger, and then transmitted to a web-based interface. The farmer managed irrigation applications. At the same time, we wanted to make sure that he was provided with sufficient information to irrigate appropriately, while maintaining statistical variability of the supplemental irrigation water applied (IW) across the field to fulfill our research purpose. In 2014, we started sending out weekly MOIST reports to the farmer. The report contained information on the soil water status and irrigation scheduling based on soil sensors and water balance calculations.

Two different methods were used to create irrigation application zones across the field: programming the two pivots (pie shape zones) and partially swapping the sprinkler nozzles (arc shape zones). Table 1 summarizes the information on the irrigation programs at each pivot. The farmer's routine irrigation schedule was 15.50 mm and 9.91 mm per revolution for the east and west pivots, respectively. The east (west) pivot panel was programmed to apply ±5.08 (±1.78) mm variation in irrigation per revolution on some pie shape zones. The control panels of pivots were Valley Select2 (Valmont Industries, Inc.) that were programmable for up to nine different pie shape zones. The program changes the irrigation rate by adjusting the pivot's travel speed, where speeding up the pivot causes less irrigation and slowing it down applies additional irrigation. Based on the pivots' characteristics and soil spatial variation, multiple banks of sprinklers were also selected and re-nozzled to form arc-shape irrigation zones. The center pivots can be operated both clockwise and counterclockwise, but were programmed only in the clockwise direction (Table 1).

Table 1. Detailed information on the supplemental irrigation programs for the two center pivots within the 73-ha supplemental irrigation field that is located in southwestern Dyer County in west Tennessee for one revolution.

	East Pivot			West Pivot		
Program Sector	Start Angle [1] (degree)	Stop Angle (degree)	Depth of Water (mm)	Start Angle (degree)	Stop Angle (degree)	Depth of water (mm)
1	90	110	10.41	275	315	9.91
2	110	0	15.49	315	335	11.68
3	0 [1]	20	20.57	335	355	8.38
4	20	40	10.41	355	235	9.91
5	40	70	15.49	235	255	11.68
6	70	90	20.57	255	275	8.38

[1] The zero degree was at north and pivots traveled clockwise.

We installed three Agspy (AquaSpy Inc., San Diego, CA, USA) soil moisture probes at three randomly selected points each year to monitor the soil water status, across pie-shape zones throughout the irrigation seasons. The AgSpy soil moisture capacitance probes were 1-m in length and obtained measurements at 10 depths at 0 to 100 cm, with 10 cm increments. The sensor output is a dimensionless number in the range 0 to 100, called the scaled frequency (*SF*), which is defined as:

$$SF = \frac{(F_a - F_s)}{(F_a - F_w)} \times 100 \tag{4}$$

where F_a is the frequency of oscillation in air (air count), F_s is the frequency of oscillation in soil (soil count), and F_w is the frequency of oscillation in water (water count). The F_a and F_w are calculated during the manufacturing of each sensor. The frequency of oscillation is related to the capacitance between sensor plates that is in turn influenced by the relative permittivity of the soil media. The relative permittivity of water is significantly greater than that of air and soil, thereby changes in soil water content will be detected by the sensor [25].

Table 2 summarizes irrigation and weather data for the 2013 and 2014 cropping seasons. The sensors were installed a couple of weeks after planting and were removed prior to the harvest period. Consequently, in situ data were not available for the whole cropping seasons. However, temperature and precipitation data from the closest weather station were obtained from the National Climate Data Center [26] to fill these gaps.

Table 2. Growing season summary of weather and supplemental irrigation data in the 73-ha study area for the 2013 and 2014 growing seasons, in comparison to the 30-year mean for these variables. The study area is located in southwestern Dyer County in west Tennessee.

Year	Variable	Month						
		May	June	July	August	September	October	November
2013	Rain, mm	23	150	190	95	79	112	63
	IW-East, mm			40	31	62		
	IW-West, mm			15	20	30		
	ET_{ref} [1], mm day^{-1}			4.33	4.43	3.92	2.49	1.28
2014	Rain, mm	143	172	56	124	120	18	
	IW-East, mm			62	31			
	IW-West, mm			20	30			
	ET_{ref} [1], mm day^{-1}	4.15	4.42	4.86	4.51	3.47	2.94	
30 year	Rain, mm	120	101	102	74	82	82	117
	Tmean, °C	21	25	27	26	22	16	10

[1] ET_{ref}: Reference evapotranspiration data that were calculated using the Turc equation (Equation (3)) from 19 July 2013 (7 May 2014) to 30 November 2013 (5 October 2014), IW: irrigation water applied by the farmer. The 30-year mean data collected from the closest weather station [26].

2.5. Multiyear Yield Data Analysis

To better understand the spatiotemporal dynamics of changes in yield, several years with different crops should also be considered [27]. Except for 2011, yield data from 2007 to 2012 (i.e., corn 2007, corn 2008, soybean 2009, cotton 2010, cotton 2012) had been collected by the producer using appropriate yield-monitor-equipped harvesters (Table 3). We combined these data with the 2013 and 2014 yield data to analyze the relative difference and temporal variance of yield on the study site under both rainfed and supplemental irrigation.

Table 3. Descriptive statistics on yield data (Mg ha^{-1}) at the field of study located in southwestern Dyer County in west Tennessee.

Year	Crop	Mean	SD
2007	Corn	7.137	4.158
2008	Corn	3.420	0.903
2009	Soybean	3.221	0.860
2010	Cotton	0.947	0.306
2012	Cotton	0.913	0.494
2013	Cotton	0.871	0.329
2014	Cotton	1.244	0.493

A multistep filtering process was designed and implemented in Microsoft Excel and ArcGIS 10.2.2 [21] to process the yield data and produce final yield maps. First, the yield maps were visually assessed using the farmer's knowledge of field conditions to identify potential unexpected patterns. Second, the data were color-coded based on harvest time to investigate the GPS tracks and movement of the harvester. Then, multiple filters were designed (e.g., using swath width, distance, speed of the harvester, change in speed) to remove outliers and erroneous data points. Last, yield data that were $>\pm 3$ standard deviations of the mean were assumed to be outliers and removed from the analysis. Then, the field was divided into 100 m^2 cells, and relative yield difference (Equation (5)) and yield temporal variance (Equation (6)) across years were calculated as follows [28]:

$$\bar{y}_i = \frac{1}{n} \sum_{k=1}^{n} \left[\frac{y_{i,k} - \bar{y}_k}{\bar{y}_k} \right] \times 100 \tag{5}$$

where n is the number of years with yield data available, $\bar{y}_i$ is the average percentage yield difference at cell i, $\bar{y}_k$ is the average yield (Mg ha^{-1}) across cells at year k, and $y_{i,k}$ is the yield value (Mg ha^{-1}) at cell i at year k.

$$\bar{\sigma}_i^2 = \frac{1}{n} \sum_{k=1}^{n} \left(y_{i,k} - \bar{y}_{i,n} \right)^2 \tag{6}$$

where $\bar{\sigma}_i^2$ is the temporal variance at cell i, $\bar{y}_{i,n}$ is the average yield across the n years, and other variables are as previously defined.

3. Results and Discussion

3.1. Field-Level Soil Heterogeneity and Application of Soil ECa

Table 4 contains descriptive statistics for the measured soil properties. The BD had its highest mean value at the deepest layer, while the mean value was almost identical among other layers. The mean water content decreased with depth, while its standard deviation slightly increased. The higher water content in the surface layer is likely attributed to textural differences among layers and also rainfall events prior to the sampling, which built the moisture level up within the top layers, but perhaps did not fully penetrate to the deeper layers. The mean sand percentage increased with depth, which was inversely proportional to a decline in silt and clay. The mean and standard deviation of the deep ECa readings (27.52 ± 18.73) were greater than those of shallow readings (24.64 ± 10.66). The standard deviation among deep ECa reading was almost twice that of shallow readings. The same result was reported by [29] on differences between the standard deviation and distribution of shallow versus deep ECa readings.

Table 4. Descriptive statistics for selected soil properties from different soil sampling layers. Soil samples were collected at four depths from 0–1 meter at 100 locations.

Variable [1]	Layer	Min.	Max.	Mean	SD
BD, g cm^{-3}	1th	1.12	1.66	1.36	0.10
	2nd	1.11	1.70	1.35	0.12
	3rd	1.06	1.86	1.34	0.12
	4th	1.17	1.78	1.40	0.13
	total	1.06	1.86	1.36	0.12
WC, %	1th	10.75	59.74	28.35	7.43
	2nd	7.27	43.12	26.02	10.78
	3rd	5.98	42.38	21.64	11.08
	4th	5.67	45.32	20.18	11.15
	total	3.94	47.61	17.94	8.49
Sand, %	1th	8.77	88.25	38.07	20.11
	2nd	0.00	94.98	46.39	31.57
	3rd	2.50	95.70	61.38	31.10
	4th	5.46	96.86	69.90	26.09
Clay, %	1th	7.37	47.56	27.55	9.04
	2nd	2.50	56.60	22.18	14.17
	3rd	1.26	47.72	14.27	11.44
	4th	0.34	37.10	11.00	7.80
Silt, %	1th	4.38	54.06	34.38	12.75
	2nd	0.00	66.51	31.43	19.85
	3rd	0.00	72.81	24.35	21.76
	4th	0.00	69.23	19.10	19.83
ECa, mS m^{-1}	shallow	1.60	48.70	24.64	10.66
ECa, mS m^{-1}	deep	1.70	162.20	27.52	18.73

[1] BD: soil bulk density, WC: soil volumetric water content at the time of sampling, ECa: apparent soil electrical conductivity, SD: standard deviation.

The soil texture drastically varied across the field such that almost the entire soil texture triangle was covered by the collected samples, except for the silt and clay textures (Figure 3). There was a shift from fine to coarse textures by depth, with sand showing the greatest particle increase. The sand had the highest absolute correlation with the soil moisture of the samples, while there was a weak negative correlation between BD and the water content (Figure 4), showing that soil texture was the dominant attribute governing water content. There was a clear pattern in clay and silt percentage plots versus water content; the majority of the samples with lower clay and silt contents belonged to the deeper layers (a cluster of black dots in the soil texture triangle), while samples from the shallower layers were more likely to have higher clay and silt contents. The opposite was seen in the sand versus water content plot.

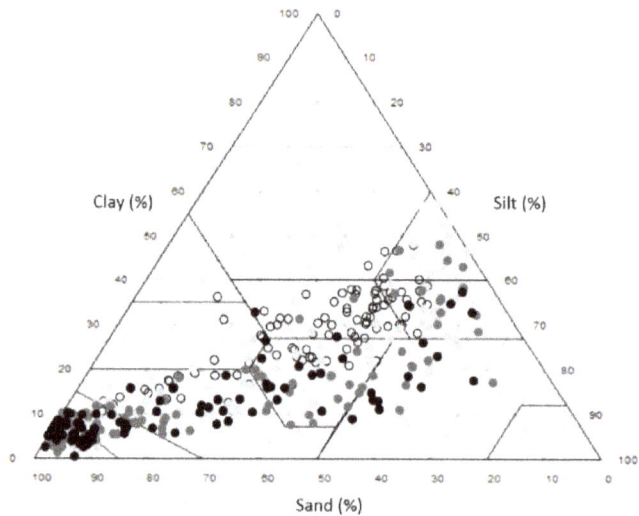

Figure 3. The textural distribution of soil samples from four different depths between 0 to 1 meter, where the darker colors correspond to the deepest depths. The samples were collected from a 73-ha two-pivot irrigation field that is located in southwestern Dyer County, Tennessee.

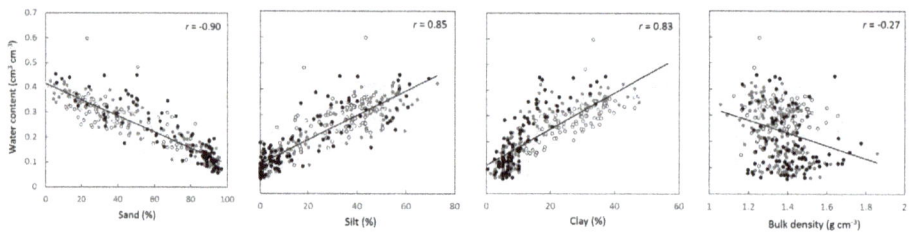

Figure 4. The relationship of 400 samples at four depths from 0–1 meter of soil texture (% Clay, Silt, & Sand) and bulk density (BD) to volumetric water content from a 73-ha two-pivot field in west Tennessee. The light to darker colors of the data markers correspond to 0–1 meter depths.

Table 5 presents the semivariogram and Global Moran's Index parameters for the selected attributes for each soil layer. The highest range did not belong to the same layer across soil properties. The average range varied from 200 m to 300 m among attributes, which was two to three times greater than the sampling intervals. The percent of nugget ranged from 18 to 50% among soil properties in the study of [30], who investigated the spatial variability of soil physical properties of alluvial soils in a 162 ha cotton field in Mississippi. This was somewhat similar to what was found for all the attributes except BD, which reached a nugget percent as high as 73 percent. The z-scores revealed all the attributes except BD within different layers had clustered patterns. BD only showed a clustered pattern at the third layer and had a random pattern at other layers.

Table 5. Semivariogram and Moran's I parameters of soil properties for different soil layers. Soil samples were collected at four depths from 0–1 meter at 100 locations.

Variable	Layer	Nugget	Sill	Range (m)	Moran's I	z-Score
* BD, g cm^{-3}	1th	0.008	0.011	526	0.087	1.181
	2nd	0.01	0.015	95	−0.086	−0.929
	3rd	0.011	0.016	280	0.137	1.802
	4th	0	0.017	100	0.091	1.221
	total	0	0.007	95	−0.007	0.038
WC, %	1th	0	44	100	0.175	2.266
	2nd	12	129	332	0.327	4.063
	3rd	0	131	206	0.284	3.545
	4th	56	125	212	0.284	3.556
	total	0	88	316	0.326	4.049
Sand, %	1th	115	446	360	0.421	5.213
	2nd	440	1119	300	0.365	4.510
	3rd	401	1037	219	0.320	3.978
	4th	413	717	260	0.300	3.747
Clay, %	1th	19	92	389	0.392	4.861
	2nd	123	215	428	0.239	3.016
	3rd	68	138	177	0.321	4.034
	4th	35	63	216	0.335	4.227
Silt, %	1th	39	174	334	0.382	4.740
	2nd	165	453	279	0.396	4.887
	3rd	211	484	200	0.270	3.366
	4th	6	10	341	0.266	3.332
ECa, mS m^{-1}	shallow	38	133	253	0.816	65.436
ECa, mS m^{-1}	deep	126	388	223	0.846	67.899

* BD: soil bulk density, WC: soil volumetric water content at the time of sampling, ECa: apparent soil electrical conductivity.

Figure 5 shows maps interpolated by ordinary kriging. The white strip expanding from the northwest to southeast of the field is a surface drainage pathway. There are three major sandy regions within the field of study at the surface layer located at: (i) surrounding pivot points at the eastern part of the field; (ii) south of the field, mostly outside of the irrigated zones; and (iii) northwest part of the field. The sequence of sand maps from the first to fourth layers illustrate how these coarse soil regions expanded across the field by depth such that sand covered the majority of the field in deeper layers. The sandy regions could be either river flood-induced sand boils or earthquake-induced sand blows. The vertical arrangement of soil textural components was not consistent across the field. The clay had its highest influence from 0–50 cm, yet sand was the dominant particle from 50–100 cm. The observed depth to sand during sampling ranged from 15–75 cm across the field, with an average depth of 40 cm for almost 40% of the sampling spots. For the rest of the samples (60%), either there was no clear immediate change from fine texture to coarse texture or sand appeared at the surface soil. The silt contributed highly in subsurface layers (25–75 cm), where it reached its highest quantity and SD (Table 4). The majority of the samples from subsurface layers (50–75 cm) with a high silt content were compacted to some extent. This compaction was also projected in relatively higher BD values from the same layers (Table 4). The BD map of the third layer corresponded well to the textural patterns, where higher BD matched coarse samples. However, it was difficult to identify a trend from the rest of the BD maps, as was expected from the results of the spatial analysis (Table 5).

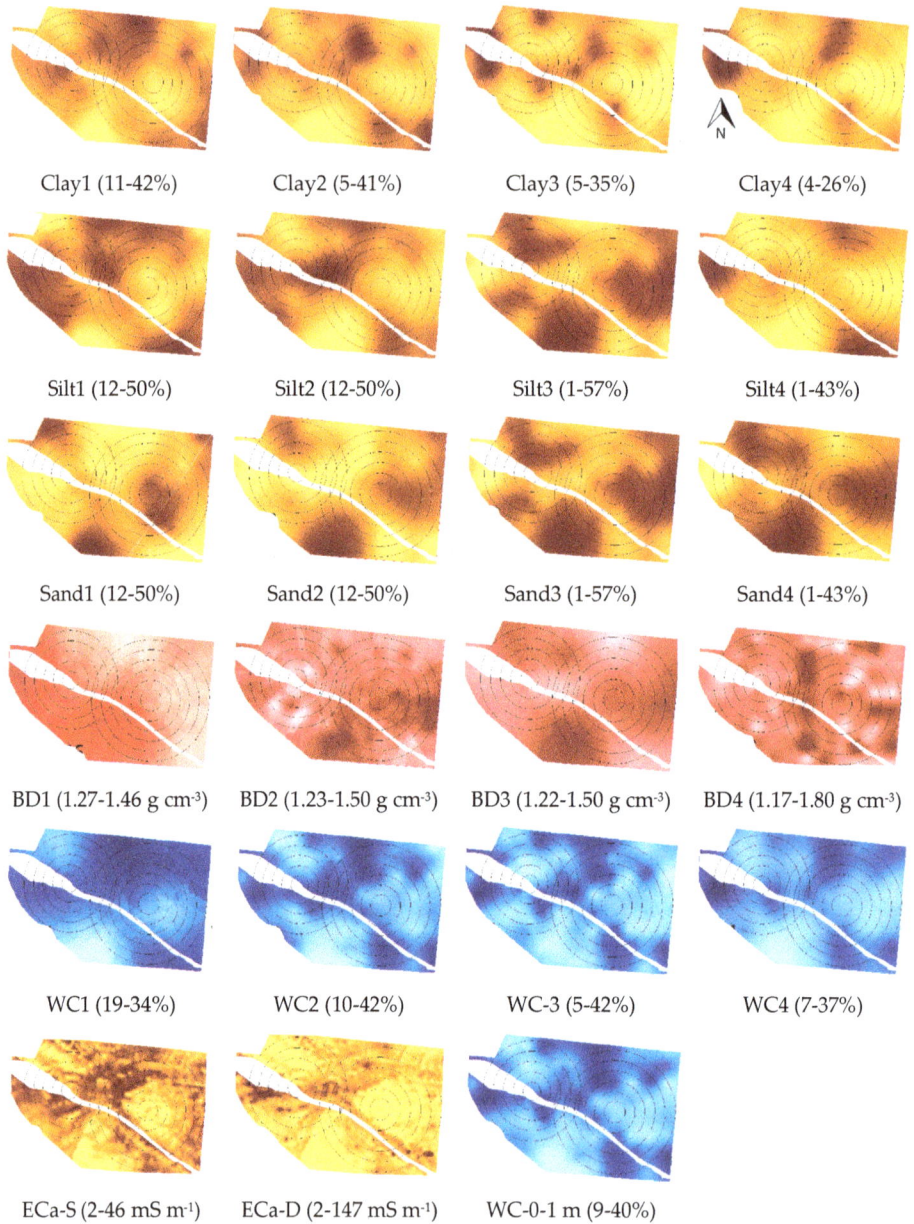

Figure 5. Maps of six different soil properties and their range of variation within a 73-ha two-pivot field in west Tennessee. These attributes include percent silt, sand, and clay, bulk density (BD, g cm^{-3}), volumetric water content (WC, %), and apparent electrical conductivity shallow (ECa-S, mS m^{-1}) and deep (ECa-D, mS m^{-1}). Numbers 1-4 denote layers 1–4 (layer 1: 0–25 cm, layer 2: 25–50 cm, layer 3: 50–75 cm, and layer 4: 75–100 cm). The maps were generated using ordinary kriging. The darker colors correspond to greater values for each attribute.

The soil water content is a dynamic property of soil with time. However, it is expected that a one-time measurement of water content values across a field provides a useful insight into the relative spatial pattern of soil hydraulic properties [31]. The water content map of the surface soil clearly showed the sandy textured areas as regions with a lower water content (Figure 5). At deeper depths, the water content maps almost exactly matched the spatial pattern of the sand maps. This occurred because coarse soils tend to dry out faster and hold less water than finer textured soils. This suggests that a one-time measurement of water content during the sampling process may be mathematically transformable to a PAW map. Overall, the water content map for the entire profile (0–100 cm) was similar to maps of individual layers.

The ECa maps tended to follow the same general spatial patterns as those for soil basic properties and water content (Figure 5). Table 6 illustrates the correlation coefficient between ECa values and soil basic data. The ECa data were moderately correlated to soil texture and water content information. The lowest correlation was between BD and ECa data. The correlation between shallow ECa and other attributes declined from layer 1 to 4, as expected, while the opposite was true for ECa deep readings. Sudduth et al. [29] showed that 90% of the shallow and deep reading responses in Veris machines were approximately obtained from the soil above the 30 cm and 100 cm depths, respectively. The sand increased with depth, hence regions with high conductivity became less pronounced in the deep ECa map as opposed to the shallow ECa map.

Table 6. Correlation coefficient between soil apparent soil electrical conductivity (ECa, mS m^{-1}) data and soil basic information at the four layers (L1–L4).

	Clay (%)				Sand (%)				Silt (%)			
	L1	L2	L3	L4	L1	L2	L3	L4	L1	L2	L3	L4
ECa-S	0.75	0.55	0.35	0.40	−0.75	−0.63	−0.45	−0.39	0.65	0.60	0.46	0.36
ECa-D	0.59	0.61	0.52	0.57	−0.62	−0.73	−0.63	−0.63	0.56	0.72	0.63	0.60

	* BD (g cm^{-3})				WC (%)			
	L1	L2	L3	L4	L1	L2	L3	L4
ECa-S	−0.01	−0.15	−0.31	−0.02	0.66	0.61	0.47	0.47
ECa-D	0.06	−0.20	−0.45	−0.13	0.63	0.71	0.64	0.65

* BD: soil bulk density, WC: soil volumetric water content at the time of sampling.

This study demonstrated that ECa is a useful surrogate for both soil texture and water content. Sudduth et al. [29] studied ECa readings on 12 fields in six states of the north-central US. They found a good relationship between ECa data and soil cation exchange capacity (CEC), as well as clay content at different times and locations, thus suggesting a general calibration equation of CEC and clay content to ECa readings. They found the most variation in ECa values in Iowa fields that had the widest range in soil texture, from loam to clay loam. In contrast, [17] reported water content as the main factor influencing ECa readings in a rainfed field, but they did not find soil texture to be a significant predictor of ECa. They found a weak correlation between water content and clay content and found that other factors including elevation and organic matter, may govern the amount of soil water content [17]. In theory, multiple factors, including the relative fractions occupied by soil, water and air, geometry and distribution of particles, and soil solution attributes, affect ECa [32]. The current introduced to measure the ECa of soil, in fact, travels through liquid, soil-liquid, and solid pathways [33]. We believe an accurate understanding of field-level soil texture variability and soil water status during the ECa measurement process is crucial to efficiently interpret ECa maps. Brevik et al. [34] studied the temporal stability of ECa data with respect to soil water content. They observed a strong influence of water content on ECa readings and found that ECa's power to differentiate soils was proportional to soil moisture. They mentioned that soil water content should be reported as an essential part of ECa studies. In this study site, the spatial heterogeneity of soil texture was the main factor that governed the spatial distribution of water content. Further studies, however, are needed to examine the efficacy of ECa for other typical field-level heterogeneity in the region, where the infiltration and redistribution of water within the root zone is governed by topography rather than soil textural variability.

3.2. Cotton Supplemental Irrigation

Table 7 summarizes the correlation coefficients between yield data and some soil properties across the field of study. To evaluate the effect of variable rate fertilizer application by the farmer on yield spatial variation, we also obtained correlation information for *P* and *K*. In general, correlation coefficients were higher for 2014 data than for 2013 data. The correlations between *P* and *K* with yield data were negligible. Given the high correlation between ECa data and PAW [18], the ECa data were used to group cotton lint yield information into four soil-based zones (Figure 6). The results illustrated in Figure 6 include irrigated areas, as well as corners of the field that were rainfed. In general, there was an increase in yield from soils with lower ECa to soils with higher ECa in 2014, but not in 2013. This is in line with the relatively higher correlation between water content/soil texture and the yield data in 2014 compared to 2013.

Table 7. Correlation coefficient values between cotton lint yield data (2013 and 2014 cropping seasons), soil properties at four depths from 0–1 m, and fertilizer application.

Layer	2013					2014				
	1	2	3	4	Total	1	2	3	4	Total
* BD, g cm^{-3}	−0.16	−0.04	−0.09	−0.07		0.00	−0.23	−0.49	−0.18	
WC, %	0.22	0.08	0.03	0.12		0.47	0.51	0.46	0.51	
Sand, %	−0.12	−0.03	−0.03	−0.08		−0.44	−0.52	−0.50	−0.53	
Clay, %	0.16	0.03	−0.03	0.01		0.40	0.44	0.42	0.47	
Silt, %	0.07	0.03	0.06	0.10		0.42	0.53	0.51	0.53	
WC33	0.18	0.06	0.05	0.12		0.40	0.50	0.52	0.51	
WC1500	0.19	0.05	0.01	0.11		0.40	0.48	0.48	0.50	
ECa-S, mS m^{-1}					0.12					0.49
ECa-D, mS m^{-1}					0.08					0.58
P, Mg ha^{-1}					−0.02					0.07
K, Mg ha^{-1}					−0.11					−0.23

* WC33 and WC1500: predicted volumetric water content at soil matric potentials −33 and −1500 kPa, respectively [18]; WC: volumetric water content at the time of sampling. Layers 1, 2, 3, and 4 were from 0–25 cm, 25–50 cm, 50–75 cm, and 75–100 cm, respectively. ECa shallow and deep readings represented approximately 0–30 cm and 0–90 cm of the soil profile, respectively.

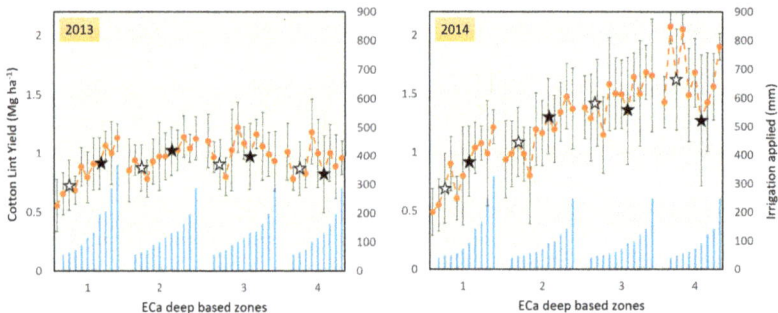

Figure 6. The effects of supplemental irrigation (blue bars, right-hand *y*-axis) on cotton lint yield (orange lines, left-hand *y*-axis) in 2013 and 2014 within a 73-ha two-pivot field in west Tennessee. The white and black five-point start symbols denote actual irrigation applications by the farmer for the west and east pivots, respectively. The deep ECa data were used to group cotton lint yield information into four zones, where ECa increases from zone 1 to zone 4, zone 1: 2–8 mS m^{-1}, zone 2: 8–18 mS m^{-1}, zone 3: 18–35 mS m^{-1}, and zone 4: 35–162 mS m^{-1}.

In 2013, only for coarse-textured soils (soils with lower ECa readings; zones 1 and 2), there was an overall positive response to supplemental irrigation. We observed no consistent positive response to higher irrigation levels for soils with higher ECa readings (i.e., zone 3 and 4 repressing fine-textured soil with higher PAW). In 2014, overall, cotton responded favorably to higher supplemental irrigation levels for the first three ECa zones, with coarse-textured soil showing the highest yield increase. The highest yield for the fine-textured zone (zone 4) did not belong to the greatest irrigation applications.

Figures 7 and 8 depict the dynamic of soil moisture (Agspy sensors) at different depths and locations over the 2013 and 2014 growing seasons, respectively. There were some missing data and bad readings mostly in 2014. In 2013, soil moisture probes *a*, *b*, and *c* were located underneath the east pivot, representing the pie shape zone with extra irrigation, pie shape zone with lower irrigation, and farmer's irrigation application, respectively. The soil moisture probes *a*, *b*, and *c* received 163 mm, 90 mm, and 133 mm seasonal IW, respectively. The average predicted PAW throughout the effective root zone (i.e., 1 m) was 0.28 m^3 m^{-3}, 0.22 m^3 m^{-3}, and 0.30 m^3 m^{-3} for probes *a*, *b*, and *c*, respectively [18]. The pattern of soil moisture was dynamic and varied among sensors at different locations and depths. At probe *a*, soil moisture depletion and replenishment occurred for sensors up to 40 cm deep during the monitoring period (i.e., days after planting (DAP): 42-133). Soil water status remained almost unchanged for deeper sensors for about 100 DAP, and then gradually exhibited a reduction, indicating that roots started to pull out water from deeper layers as the ET demand increased. At probe *b*, rainfall plus irrigation kept the soil moisture at a fairly constant level up to about 80 DAP for all sensors, while fluctuations decreased by depth, as expected. After that, there was a substantial depletion in soil moisture for sensors up to 50 cm, which even expanded to deeper sensors at about 95 DAP. At probe *c*, the overall trend was similar to that of probe *a*. Toward the end of the growing season, much rainfall at 112 DAP occurred that refilled the shallow layers for all soil moisture probes and also penetrated to deeper layers such that there was an increase in readings by the soil moisture sensors at 30 and 40 cm and no decrease for deeper sensors up to the end of the monitoring period.

In 2014, soil moisture probes *a*, *b*, and *c* were mainly located underneath the west pivot, representing the central area irrigated by both pivots, pie shape zone with lower irrigation, and farmer's irrigation application, respectively. Soil moisture probes *a*, *b*, and *c* (Figure 8) received 142, 42, and 50 mm of IW, respectively, during the 2014 cropping season. Within the effective root zone (i.e., 1m), the average PAW values predicted for probes *a*, *b*, and *c* were 0.19 m^3 m^{-3}, 0.33 m^3 m^{-3}, and 0.23 m^3 m^{-3}, respectively [18]. Unlike 2013, most of the deeper sensors in 2014 showed some fluctuations starting at about 70 DAP, meaning that the crop started to use water from deeper layers as the crop water requirement increased. We attribute this to (i) bigger plants with larger canopies, and hence a higher ET demand; (ii) lower irrigation in 2014 compared to 2013; and (iii) lower irrigation under the west pivot (where we had sensors installed in 2014) compared to the east pivot (where we had sensors installed in 2013). Trends in probes *b* and *c* were similar. There were more fluctuations in the shallow sensors in probe *a*, since this sensor was irrigated by both pivots and was located on a coarse-textured soil with a low PAW. In both years, heavy rainfall events were responsible for big changes in the soil water status within the soil profile and considering sensor fluctuations, they usually penetrated deep down to 50 cm. Irrigation events, however, mostly refilled shallow layers up to 20 cm and barley influenced sensors deeper than 30 cm.

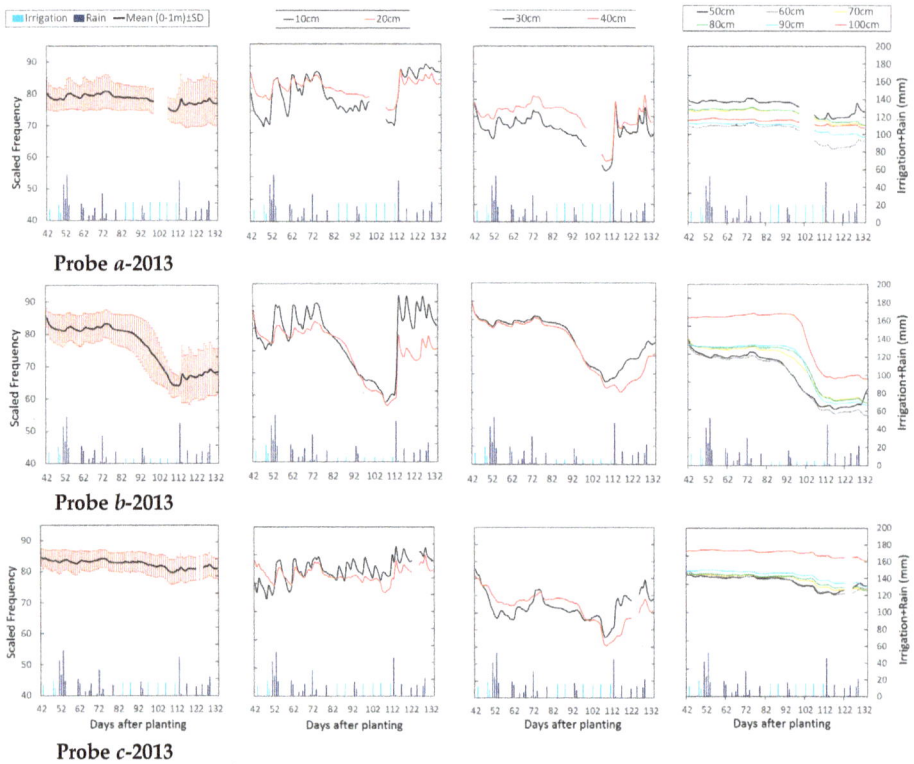

Figure 7. The change in soil moisture, measured as scale frequency (SF, Equation (4)), from 42 days after planting throughout the 2013 growing season in a 73-ha two-pivot supplemental irrigation field in west Tennessee. Light and dark blue bars show irrigation and rainfall, respectively. SF was measured using Agspy soil sensor probes at different locations and 10 depths from 0–1 m. In 2013, soil moisture probes *a*, *b*, and *c* were located underneath the east pivot, representing the pie shape zone with extra irrigation, pie shape zone with lower irrigation, and farmer's irrigation application, respectively.

The cotton lint response to supplemental irrigation differed across soil types. For soils with a lower PAW, there was a positive response to irrigation in comparison to rainfed, where soil moisture deficit is expected to reduce the boll number and yield [9]. The cotton response to irrigation was not consistent for soils with a higher PAW, except that a yield reduction occurred underneath both pivots for high irrigation rates in both cropping seasons. This is in line with the reported results in the literature, indicating that under wet conditions, excessive irrigation decreased the yield of cotton lint [11,12]. In 2013, the cotton lint yield was only 12% higher where we placed probe *a* (IW = 163 mm, predicted PAW = 0.28 $m^3 \ m^{-3}$) in comparison to the yield at probe *b* (IW = 90 mm, predicted PAW = 0.22 $m^3 \ m^{-3}$), even though there was a remarkable difference in soil water status throughout the growing season between the two soil moisture probes (Figure 7). On the other hand, in 2014, the yield difference between the exact same spots with a similar relative difference in IW increased by 44%. We attribute this to the wet season and delayed planting in 2013, which significantly affected the cotton response to irrigation. Delayed planting influences heat unit accumulation and distribution, which was underscored as an important factor for the short season cotton response to supplemental irrigation by [13]. Moreover, irrigation is expected to increase the number of bolls, but delays cutout (i.e., cessation of flowering) since irrigation continues the vegetative growth for a longer period. We believe that rapid canopy expansion occurred on soil with a higher PAW in 2013 due to excessive water within the crop effective root zone.

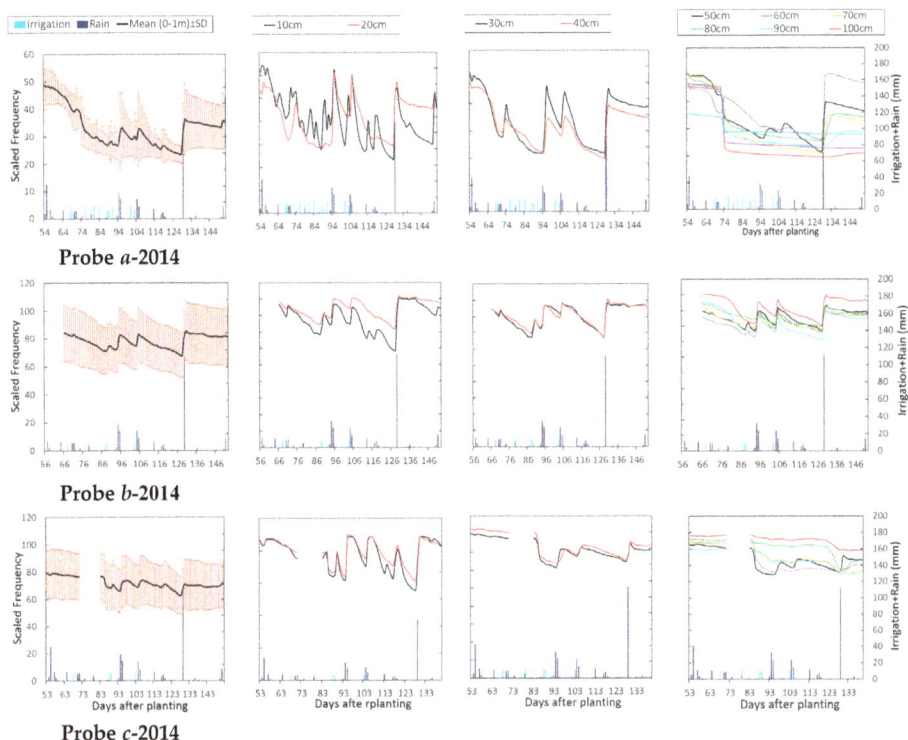

Figure 8. The change in soil moisture capacitance, measured as scale frequency (SF, Equation (4)), from 53 days after planting throughout the 2014 growing season in a 73-ha two-pivot supplemental irrigation field in west Tennessee. Light and dark blue bars show irrigation and rainfall, respectively. SF was measured using Agspy soil sensor probes at different locations and 10 depths from 0–1 m. In 2014, soil moisture probes *a*, *b*, and *c* were mainly located underneath the west pivot, representing the central area irrigated by both pivots, pie shape zone with lower irrigation, and farmer's irrigation application, respectively.

Both 2013 and 2014 were relatively wet years. In fact, the rainfall was always above the long term average, except for July 2014. There were some heavy rainfall events during the growing seasons, which caused a significant increase in the soil water content. This is a typical situation in west Tennessee, with unexpected rainfall events, where temporal changes in rainfall patterns significantly affect the yield response to supplemental irrigation across years. Bajwa and Vories [11] reported the same complexity on cotton irrigation scheduling in a moderately humid area in Arkansas when rainfall was plentiful and caused a yield reduction for high irrigated crops. Monitoring the soil water status revealed that rainfall events refilled the top soil and penetrated into deeper layers, while supplemental irrigation mostly influenced the shallow layer up to 20 cm. Therefore, any sustainable irrigation management in this region should take effective rainfall into account for site-specific irrigation scheduling. Sensors indicated fast depletion for soils with lower PAWs. This caused the crop to start using water from deeper layers as the cropping season advanced and ET demand increased. The sensor located on the overlap region of the two pivots showed that more frequent irrigation could prevent the shallow soil layer from substantial depletion and possible yield reduction due to water stress and thus should be considered as a potential irrigation strategy for coarse-textured soils with low PAWs throughout the study site. On the other hand, days with no rainfall and irrigation could be beneficial for soils with high PAWs since cotton usually responds favorably to periods of water stress adequate to reduce vegetation expansion. The farmer's goal was to tailor irrigation decisions to dominant soil areas, while avoiding over and under irrigation as much as possible. The results, however, show that uniform irrigation management can be detrimental to the field's overall production and water use efficiency. Consequently, variable rate irrigation strategies were predicted to enhance cotton production compared to current uniform irrigation management practiced by the grower [35]. There may be a requirement to practice a dynamic zoning strategy considering available water for the crop within its effective root zone during growing seasons [36]. Soil moisture data will be instrumental to making the most informed decisions for each soil and to determining how much different soils of a variable texture need to be irrigated to maximize production.

3.3. Multiyear Yield Analysis

Figure 9 illustrates the cotton yield maps for 2012, 2013, and 2014; mean yield map (from Equation (5)); and standard deviation yield map (from Equation (6)). We included the 2012 yield map, a year before this experiment when uniform irrigation had been applied by the farmer, to better represent the effect of soil spatial variation on the cotton yield. Thematic cotton yield maps for 2012 and 2014 cropping seasons followed the same spatial distribution as soil texture and water content maps, but the 2013 yield map showed a different pattern. This finding agrees with the low correlation observed between cotton yield data in 2013 and soil properties (Table 7). The spatial analysis of the multiyear mean yield map (ranged from −79 to 98) showed substantial similarity to the PAW map developed for the field of study by [18]. There were three regions with a lower yield and all on coarse-textured soils with lower PAW. The highest yield temporal stability also belonged to the regions with low PAW, (i) in the southern part of the field outside pivots coverage and (ii) in an area surrounding the east pivot point. The yield temporal stability was lower for other parts of the field, but it was hard to identify any cluster of cells with a similar temporal variance. We mainly attribute this to different rainfall patterns and irrigation regimes across years, and their effects on the yield across soil types, which caused substantial mean yield variability across years (Table 3). For instance, for cotton and corn, there was as much as 43% and 110% temporal differences between mean yields across years, respectively. However, it is known that other attributes related to management, water, crop, and climate affect the crop yield in a complex manner and the crop yield maps per se only provide limited information about the influence of each attribute [31]. In 2012 and 2014, the mean yield and standard deviation were higher than those for 2013, indicating a decline in yield on soils with higher available water in 2013 due to the delayed planting and excessive rainfall throughout the growing season.

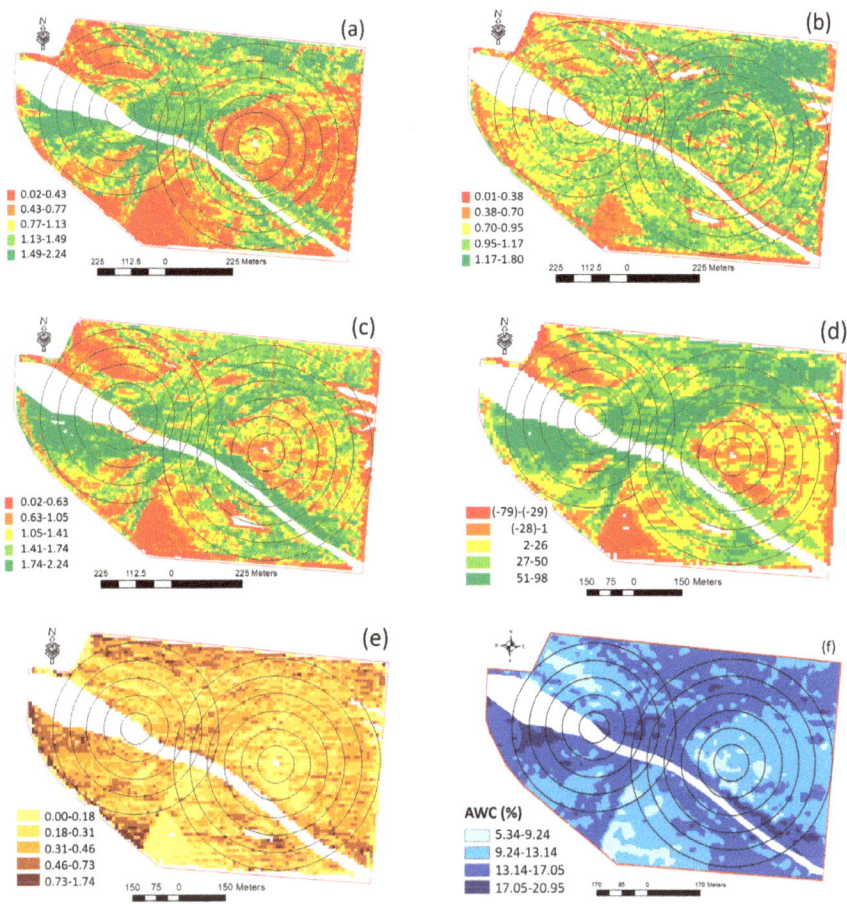

Figure 9. The cotton lint yield map (Mg ha^{-1}) time series from 2012 to 2014 (**a–c**), plus mean ((**d**), using Equation (5)) and standard deviation cotton yield maps ((**e**), using Equation (6)) that were derived from the producer's harvester data (Table 3, seven years of data for cotton, corn and soybean) for a 73-ha field in west Tennessee. The map of plant available water content (**f**) within the crop effective root zone (i.e., 100 cm) is adapted from the study by [18].

4. Conclusions

Irrigation investment has been expanding across the humid areas of the US Cotton Belt for the last 20 years because of the stabilization of yields and high commodity prices. Recent advances in modern instrumentation and measurement techniques, such as on-the-go sensing, remote sensing, and wireless networks of sensors, make site-specific on-farm experimentation possible for farmers. This is essential for field-level cotton irrigation management in humid areas as a complex problem due to substantial spatiotemporal heterogeneity in soil and weather-related parameters. In this study, we used a variety of information that is relatively easily collected by farmers, to investigate the impact of the spatial and temporal heterogeneity of soil attributes, including field-collected soil texture, moisture, ECa, and bulk density, on the cotton response to supplemental irrigation. We found that ECa was a useful proximal attribute to understand field-level spatial variability of alluvial soils in the region. Analyzing crop yield against maps of soil characteristics revealed that soil texture and soil water content remarkably influenced yield patterns, suggesting that variable rate irrigation is the appropriate irrigation scenario for this mixture of soils. We also found that other factors, including cropping season length and in-season rainfall pattern, may change or even reverse the expected lint yield from an irrigation treatment for a specific soil type. While soil variation is inherent and not controlled by farmers, irrigation, if well-scheduled, could be the key factor to optimizing the crop production and water use efficiency. The use of soil moisture sensors can help monitor soil water status and adjust irrigation application based on in-season rainfall patterns and within-field soil variability.

Author Contributions: Conceptualization, A.H. and B.L.; methodology, A.H. and B.L.; software, A.H. and B.L.; validation, A.H. and B.L.; formal analysis, A.H.; investigation, A.H.; resources, B.L..; Data curation, A.H., B.L., W.C.W., S.G., T.G., M.Z., and P.V.; writing—original draft preparation, A.H.; writing—review and editing, A.H., B.L., R.W-A., and T.G.; visualization, A.H.; supervision, B.L.; funding acquisition, B.L.

Funding: This project was funded in part by Cotton Inc. and the United States Department of Agriculture (USDA) Natural Resources Conservation Service (NRCS) Conservation Innovation Grants (CIG).

Acknowledgments: We appreciate Pugh Brothers Farms for allowing us to implement this project on their property.

Conflicts of Interest: The authors declare no conflict of interest. The funders had no role in the design of the study; in the collection, analyses, or interpretation of data; in the writing of the manuscript, or in the decision to publish the results.

References

1. FAO (Food and Agriculture Organization of the United Nations). *Statistical Yearbook 2013: World Food and Agriculture*; FAO: Rome, Italy, 2013.
2. NASS (National Agricultural Statistics Service). 2007 Census of Agriculture: Farm and Ranch Irrigation Survey, National Agricultural Statistics Service, USDA. 2010. Available online: http://www.agcensus.usda.gov/index.php (accessed on 22 March 2019).
3. Tennessee Farm Bureau Federation. Irrigation: Solving Potential Challenges: Policy Development 2013. 2013. Available online: https://www.tnfarmbureau.org/wp-content/uploads/2010/10/Irrigation.pdf (accessed on 22 March 2019).
4. Wong, M.T.F.; Asseng, S. Determining the Causes of Spatial and Temporal Variability of Wheat Yields at Sub-field Scale Using a New Method of Upscaling a Crop Model. *Plant Soil* **2006**, *283*, 203–215. [CrossRef]
5. Guo, W.; Maas, S.J.; Bronson, K.F. Relationship between cotton yield and soil electrical conductivity, topography, and Landsat imagery. *Precis. Agric.* **2012**, *13*, 678–692. [CrossRef]
6. Graveel, J.G.; Fribourg, H.A.; Overton, J.R.; Bell, F.F.; Sanders, W.L. Response of Corn to Soil Variation in West Tennessee, 1957–1980. *JPA* **1989**, *2*, 300–305. [CrossRef]
7. Perry, C.; Barns, E. *Cotton Irrigation Management for Humid Regions*; Cotton Inc.: Cary, NC, USA, 2012.
8. Vellidis, G.; Liakos, V.; Perry, C.; Tucker, M.; Collins, G.; Snider, J.; Andreis, J.; Migliaccio, K.; Fraisse, C.; Morgan, K.; et al. A smartphone app for scheduling irrigation on cotton. In Proceedings of the 2014 Beltwide Cotton Conference, New Orleans, LA, USA, 6–8 January 2014; Boyd, S., Huffman, M., Robertson, B., Eds.; National Cotton Council: Memphis, TN, USA, 2014.
9. Pettigrew, W.T. Moisture Deficit Effects on Cotton Lint Yield, Yield Components, and Boll Distribution. *Agron. J.* **2004**, *96*, 377. [CrossRef]
10. Suleiman, A.A.; Soler, C.M.T.; Hoogenboom, G. Evaluation of FAO-56 crop coefficient procedures for deficit irrigation management of cotton in a humid climate. *Agric. Water Manag.* **2007**, *91*, 33–42. [CrossRef]
11. Bajwa, S.G.; Vories, E.D. Spatial analysis of cotton (*Gossypium hirsutum* L.) canopy responses to irrigation in a moderately humid area. *Irrig. Sci.* **2007**, *25*, 429–441. [CrossRef]
12. Bronson, K.F.; Booker, J.D.; Bordovsky, J.P.; Keeling, J.W.; Wheeler, T.A.; Boman, R.K.; Parajulee, M.N.; Segarra, E.; Nichols, R.L. Site-Specific Irrigation and Nitrogen Management for Cotton Production in the Southern High Plains. *Agron. J.* **2006**, *98*, 212. [CrossRef]
13. Gwathmey, C.O.; Leib, B.G.; Main, C.L. Lint yield and crop maturity responses to irrigation in a short-season environment. *J. Cotton Sci.* **2011**, *15*, 1–10.
14. Grant, T.J.; Leib, B.G.; Duncan, H.A.; Main, C.L.; Verbree, D.A. A deficit irrigation trial in differing soils used to evaluate cotton irrigation scheduling for the Mid-South. *J. Cotton Sci.* **2017**, *21*, 265–274.

15. Khosla, R.; Westfall, D.G.; Reich, R.M.; Mahal, J.S.; Gangloff, W.J. Spatial Variation and Site-Specific Management Zones. In *Geostatistical Applications for Precision Agriculture*; Springer Nature: Berlin, Germany, 2010; pp. 195–219.

16. Gooley, L.; Huang, J.; Page, D.; Triantafilis, J. Digital soil mapping of available water content using proximal and remotely sensed data. *Soil Use Manag.* **2013**, *30*, 139–151. [CrossRef]

17. McCutcheon, M.C.; Farahani, H.J.; Stednick, J.D.; Buchleiter, G.W.; Green, T.R. Effect of Soil Water on Apparent Soil Electrical Conductivity and Texture Relationships in a Dryland Field. *Biosyst. Eng.* **2006**, *94*, 19–32. [CrossRef]

18. Haghverdi, A.; Leib, B.G.; Washington-Allen, R.A.; Ayers, P.D.; Buschermohle, M.J. High-resolution prediction of soil available water content within the crop root zone. *J. Hydrol.* **2015**, *530*, 167–179. [CrossRef]

19. Blake, G.R.; Hartge, K.H. Bulk density. In *Methods of Soil Analysis. Part 1*, 2nd ed.; Agron. Monogr. 9; Klute, A., Ed.; ASA and SSSA: Madison, WI, USA, 1986; pp. 363–375.

20. Cressie, N. Spatial prediction and ordinary kriging. *Math. Geol.* **1988**, *20*, 405–421. [CrossRef]

21. ESRI. *ArcMap (Version 10.2.2)*; Environmental Systems Resource Institute: Redlands, CA, USA, 2014.

22. Getis, A.; Ord, J.K. The Analysis of Spatial Association by Use of Distance Statistics. *Geogr. Anal.* **2010**, *24*, 189–206. [CrossRef]

23. Turc, L. Estimation of irrigation water requirements, potential evapotranspiration: A simple climatic formula evolved up to date. *Ann. Agron.* **1961**, *12*, 13–14.

24. Lu, J.; Sun, G.; McNulty, S.G.; Amatya, D.M. A comparison of six potential evapotranspiration methods for regional use in the southeastern United States. *J. Am. Water Resour. Assoc.* **2005**, *41*, 621–633. [CrossRef]

25. Paterson, N.D.; Richard, J.C.; Neil, M.W. Soil Moisture Sensor with Data Transmitter. U.S. Patent 12/310,946, 10 December 2009.

26. National Climate Data Center. National Climate Data Center Home Page. 2015. Available online: http://www.ncdc.noaa.gov/data-access/land-based-station-data (accessed on 13 June 2015).

27. Joernsgaard, B.; Halmoe, S. Intra-field yield variation over crops and years. *Eur. J. Agron.* **2003**, *19*, 23–33. [CrossRef]

28. Basso, B.; Bertocco, M.; Sartori, L.; Martin, E.C. Analyzing the effects of climate variability on spatial pattern of yield in a maize–wheat–soybean rotation. *Eur. J. Agron.* **2007**, *26*, 82–91. [CrossRef]

29. Sudduth, K.; Kitchen, N.; Wiebold, W.; Batchelor, W.; Bollero, G.; Bullock, D.; Clay, D.; Palm, H.; Pierce, F.; Schuler, R.; et al. Relating apparent electrical conductivity to soil properties across the north-central USA. *Comput. Electron. Agric.* **2005**, *46*, 263–283. [CrossRef]

30. Iqbal, J.; Thomasson, J.A.; Jenkins, J.N.; Owens, P.R.; Whisler, F.D. Spatial Variability Analysis of Soil Physical Properties of Alluvial Soils. *Soil Sci. Soc. Am. J.* **2005**, *69*, 1338. [CrossRef]

31. Corwin, D.L.; Lesch, S.M.; Shouse, P.J.; Ayars, J.E.; Soppe, R.; Ayars, J.E. Identifying Soil Properties that Influence Cotton Yield Using Soil Sampling Directed by Apparent Soil Electrical Conductivity. *Agron. J.* **2003**, *95*, 352–364. [CrossRef]

32. Friedman, S.P. Soil properties influencing apparent electrical conductivity: A review. *Comput. Electron. Agric.* **2005**, *46*, 45–70. [CrossRef]

33. Rhoades, J.D.; Corwin, D.L.; Lesch, S.M. Geospatial measurements of soil electrical conductivity to assess soil salinity and diffuse salt loading from irrigation. In *Solar Eruptions and Energetic Particles*; American Geophysical Union (AGU): Washington, DC, USA, 1999; Volume 108, pp. 197–215.

34. Brevik, E.C.; Fenton, T.E.; Lazari, A. Soil electrical conductivity as a function of soil water content and implications for soil mapping. *Precis. Agric.* **2006**, *7*, 393–404. [CrossRef]

35. Haghverdi, A.; Leib, B.G.; Washington-Allen, R.A.; Buschermohle, M.J.; Ayers, P.D. Studying uniform and variable rate center pivot irrigation strategies with the aid of site-specific water production functions. *Comput. Electron. Agric.* **2016**, *123*, 327–340. [CrossRef]

36. Haghverdi, A.; Leib, B.G.; Washington-Allen, R.A.; Ayers, P.D.; Buschermohle, M.J. Perspectives on delineating management zones for variable rate irrigation. *Comput. Electron. Agric.* **2015**, *117*, 154–167. [CrossRef]

agriculture

MDPI

Article

Calibration and Global Sensitivity Analysis for a Salinity Model Used in Evaluating Fields Irrigated with Treated Wastewater in the Salinas Valley

Prudentia Zikalala [1],*, Isaya Kisekka [2] and Mark Grismer [2]

[1] Department of Land, Air, and Water Resources, University of California, One Shield Avenue, Davis, CA 95616, USA
[2] Department of Land, Air, and Water Resources & Biological and Agricultural Engineering, University of California, One Shield Avenue, Davis, CA 95616, USA; ikisekka@ucdavis.edu (I.K.); megrismer@ucdavis.edu (M.G.)
* Correspondence: pgzikalala@ucdavis.edu; Tel.: +1-530-220-3714

Received: 9 January 2019; Accepted: 25 January 2019; Published: 1 February 2019

Abstract: Treated wastewater irrigation began two decades ago in the Salinas Valley of California and provides a unique opportunity to evaluate the long-term effects of this strategy on soil salinization. We used data from a long-term field experiment that included application of a range of blended water salinity on vegetables, strawberries and artichoke crops using surface and pressurized irrigation systems to calibrate and validate a root zone salinity model. We first applied the method of Morris to screen model parameters that have negligible influence on the output (soil-water electrical conductivity (EC_{sw})), and then the variance-based method of Sobol to select parameter values and complete model calibration and validation. While model simulations successfully captured long-term trends in soil salinity, model predictions underestimated EC_{sw} for high EC_{sw} samples. The model prediction error for the validation case ranged from 2.6% to 39%. The degree of soil salinization due to continuous application of water with electrical conductivity (EC_w) of 0.57 dS/m to 1.76 dS/m depends on multiple factors; EC_w and actual crop evapotranspiration had a positive effect on EC_{sw}, while rainfall amounts and fallow had a negative effect. A 50-year simulation indicated that soil water equilibrium ($EC_{sw} \leq 2dS/m$, the initial EC_{sw}) was reached after 8 to 14 years for vegetable crops irrigated with EC_w of 0.95 to 1.76. Annual salt output loads for the 50-year simulation with runoff was a magnitude greater (from 305 to 1028 kg/ha/year) than that in deep percolation (up to 64 kg/ha/year). However, for all sites throughout the 50-year simulation, seasonal root zone salinity (saturated paste extract) did not exceed thresholds for salt tolerance for the selected crop rotations for the range of blended applied water salinities.

Keywords: treated wastewater irrigation; salinization; model simulation; global sensitivity analysis

---

1. Introduction

Salinization of soils, groundwater, and surface waters from irrigation is a well-known problem often associated with the decline of ancient civilizations dependent on irrigated agriculture around the world, such as Mesopotamia [1]. Today, the salinity problem associated with irrigation in low rainfall regions continues to have numerous grave economic, social, and political consequences. For example, there is a high economic cost associated with salinity; the US Bureau of Reclamation spends $32 million annually to limit salt additions to the Colorado River and the Natural Resource Natural Resource Conservation Service-US Department of Agriculture (USDA-NRCS) spends some $13 million annually to control salinity in irrigation programs across the upper Colorado River Basin [2]. Simultaneously, as competition for available freshwater resources intensifies and use of treated wastewater (recycled water) having greater salinity grows to meet agricultural water demands, a key question inevitably remains, is term-term use of recycled water sustainable? In the 1980s, discussions about "sustainable agriculture" raised questions about changes in soil quality. Soil-water salinity is a transient condition whereby salts concentrate in the soil following root water uptake by plants as well as water loss by evaporation at the soil surface. Subsequent irrigation or rainfall can dilute the soil-water salinity or the solutes can be removed from the system by leaching to subsurface drains, or through deep percolation below the root zone. In areas having fine-textured soils overlying a shallow water table, additional root zone salinization can occur through capillary rise from the saline water table [3–5]. Salinity risks also increase when saline water is used for irrigation and when poor fertilizer and poor irrigation management are combined.

Salinity hazards caused by irrigation depend on the type of salts, soil, and climatic conditions, crop species, and the amount, quality, and frequency of water applications [6]. Increased irrigation efficiency through adoption of advanced irrigation technologies such as micro-irrigation and sprinkler systems may result in less water used in fields but may also decrease the leaching required to maintain satisfactory root zone salinity during the growing season. While advanced irrigation technologies are beneficial for increasing water productivity and protecting groundwater resources from pollutant leaching, the low leaching fractions may lead to soil salinization. In addition, surface runoff pickup of salts and leaching enable accumulated field salts to degrade river and groundwater resources. These trade-offs also suggest that refined guidelines for use of treated wastewater for irrigation are needed and could be aided through root zone salinity modeling.

Groundwater salinization is occurring in aquifers along the California coast [7] and is especially critical in the Salinas Valley of the Central Coast as seawater intrusion threatens groundwater supplies critical for irrigation of high-value fruit and vegetable crops. As a means to limit seawater intrusion, tertiary treated wastewater was made available for agricultural use in the Salinas Valley since 1998 as an alternative or supplement to groundwater and concerns are growing about possible root zone salinization in fields receiving recycled water. Accumulation of salinity in the crop root zone progressively decreases yields. For example, during a 13-year field experiment in the Castroville area, Platts and Grismer [8] observed an upward trend in soil electrical conductivity (EC) and chloride (Cl) indicating a soil salinization threat and a possible growing Cl toxicity threat to crop production in the Salinas Valley. The range of increase in EC in the root zone for sites irrigated with blended well and treated wastewater was 18 to 63% and an increase in Cl ranged from 48 to 510%. Moreover, agricultural return flows account for an estimated 33% of annual recharge to groundwater in the Valley [9]. In a geochemical analysis, Vengosh and others [10] suggested that 3–10 mm/year of vertical seepage associated with agriculture adversely affects the Valley's groundwater quality. On the other hand, Platts and Grismer [11] found that annual winter rainfall of roughly 250 mm was required to adequately leach accumulated salts associated with recycled water use for irrigation in the Valley. Moreover, from the lower Salinas River at Gonzales to the estuary, salinity is listed under EPA 303d indicating that salinity in the Salinas Valley threatens sensitive surface water supply and ecosystems.

Root zone soil-water models have been developed in an effort to gain both an understanding of the complex processes associated with soil–water–chemistry dynamics in the root zone and to provide guidance for water managers and growers. Dynamic soil-water models quantify many physical-chemical-biological interactions in irrigated agricultural systems and enable predictions to assess spatio-temporal changes in soil salinity during and between growing seasons [12]. Soil EC is one of nineteen measures advocated by the Soil Health Institute [13] as a measure of agricultural sustainability and is a critical output parameter from these models. Further, Maas and Hoffman [14] and Rhoades and others [15] developed crop-threshold EC values to assure successful use of saline water for irrigation. Important factors in such models include daily rain depths and evapotranspiration (ET) demands, soil properties, crops and irrigation method, water application depths and quality, and chemical factors such as salt precipitation and dissolution rates within the root zone. Understanding and predicting how root zone salinity changes in time under different irrigation methods and cropping systems provides insight into possible groundwater and surface water salinization. Several models have been developed to estimate the soil water balance of the soil–plant-atmosphere system, Decision Support System for Agrotechnology Transfer (DSSAT) for example is best suited to simulate process-based crop growth and development; the model does not include a salinity module and this model uses a "tipping bucket" water balance approach for soil hydrologic and water redistribution processes [16]. HYDRUS-1D simulates water flow in soils using the numerical solutions of the Richards equation, however, its simulation of crop-related processes is limited. Moreover, for long-term, multi-cropping simulations, HYDRUS-1D requires loose coupling with an external crop model for estimations of evapotranspiration. As such, we elected to use a simple daily time-step soil salinity model based on soil-water storage in four rootzone quarters and applicable to long records of meteorological data. This enables us to take into account a number of site-specific factors including soil properties, rainfall patterns, crop type, and irrigation methods to establish the effect of these factors on long-term soil salinity.

While dynamic soil-water processes can be simulated at multiple time scales, a daily time step is typically deployed because it represents the time scale at which rainfall, water application, and ET information is more readily available and because many of the actual root zone processes occur within hours to a day timeframe. While [11] employed a daily soil-water balance model to examine the effects of various hydrologic factors on soil salinity over the 13-year study, they did not include root zone soil-water chemistry, upward flow from shallow water tables and fertilizer management processes. Isidoro and Grattan [17] developed a root zone soil salinity and water balance model with root-water-extraction assumptions similar to foundational models of the past [18]. We extend this model to further account for drainage under shallow groundwater conditions to enable inclusion of saline water table effects on root zone soil water and salinity as similarly described in Reference [19].

Due in part to the greater capillary rise in finer-textured soils, greater upward flow rates from shallow water tables has been found in loamy soils than in sandy soils having small capillary rise, or in clayey soils with very slow permeability [20]. Crop water use from shallow water tables is controlled by its depth and water quality, crop type, growth stage and salt tolerance, and water application frequency and depth as affected by the vadose zone hydraulic conductivity [4,5,19,21–25]. In the Salinas Valley study area, groundwater depths range from 7.5 to 11.6 m (24.4 to 38 feet) below ground surface with maximum groundwater depths occurring in the fall following the summer irrigation season. In the 13-year study by Reference [8], they noted that observed Cl accumulation in the soil profile may have resulted from upward flow of saline groundwater.

Any modeling effort is a representation that necessarily simplifies reality; however, simulations enable investigation of "what if?" questions. Previous analyses by References [8,11] of overall soil salinity changes and leaching during the 13-year field study suggested that root zone salinity levels could be managed by winter rains when irrigating with blended recycled water. However, they underscored that more detailed process analyses were required to better elucidate what applied water salinity levels were tolerable. Here, we seek to quantify (model) long-term (50 years) trends in soil salinity within the Castroville area of the Salinas Valley when shallow groundwater is present as affected by irrigation with recycled water of varying salinity. Further, model applicability to a region, requires model calibration and validation for the study-site conditions so we explore use of a new two-step process that first identifies the critical model parameters and then focuses on those to calibrate the model. Use of models enables a comparatively inexpensive and environmentally-safe technique to evaluate the long-term effects of various agricultural management scenarios on soil salinity while also providing an aid for water managers considering these complex processes.

We use observations from the 13-year field experiment evaluating the long-term effects from use of varying fractions of recycled water (i.e., salinity) for vegetable and strawberry production on soil salinity in the Castroville area building on the previous efforts by Platts and Grismer [8,11]. Study objectives were to:

- Perform global sensitivity analysis of the modified Isidoro and Grattan [17] root zone salinity model to find the parameters most sensitive to model outputs:

- Complete a model calibration and validation using parameters to which model outputs are most sensitive as a guide: and
- Predict long-term (five decades) root zone salinity, salt output load with deep percolation and salt output load with surface runoff from fields using treated wastewater for irrigation.

2. Materials and Methods

2.1. Study Area

The study area in Castroville, overlies two main aquifers referred to as the "180-foot" and "400-foot" aquifers, respectively. These were formed from fluvial sands and gravels associated with the old Salinas River channel and possible delta conditions. Above and between the two aquifers are deposits of blue clay overlying the "180-foot" aquifer range from 8-m thick at Salinas to more than 30-m thick at Castroville [26–28]. The typical overlying soil profile in the study area is comprised of Pacheco sandy to clay loams as summarized in Table 1.

Table 1. Average soil profile texture variations in the study region [29].

Texture	Depth (m)	Textural Fractions (%)				Conductivity K_s (mm/d)	Bulk Density (kg/m^3)
		Sand	Clay	Silt	OM		
Clay loam	0–0.6	20–45	27–35	20–53	2	121–363	1660
Sandy loam	0.6–0.9	35–70	15–27	3–44	0.5	363–1218	1640
Loam	0.9–1.2	30–50	20–27	28–50	0.5	363–1218	1640
Silty clay loam	1.2–3	15–20	27–35	45–58	0.5	36–121	1700

We assembled the base data for the model using the estimates for saturated hydraulic conductivity, organic matter content, soil texture, and bulk density from SoilWeb [29] an interactive webtool to access detailed soil survey data (SSURGO). We then determined saturation water content (Ts), wilting point (Twp), and field capacity (Tfc) according to soil texture using artificial neural networks techniques implemented in Hydrus-1D [30]. Meteorological records for 1983 to 2014 were taken from the local (California Irrigation Management Information System (CIMIS) station number 19 [31] and the average monthly rainfall and grass-reference ET_o are shown in Figure 1. Average monthly reference ET_o exceeded average monthly rainfall during April through November, annual ET_o ranged from 862.3 mm to 1072.6 mm ($\pm$5.3%). Rainfall was concentrated from November to March (87% of annual rainfall) with annual rainfall depths ranging from 134.5 mm to 1026 mm ($\pm$47.1%) as shown in Figure 2.

The main crops grown in the project area are cool-season vegetables (lettuce, broccoli, cauliflower, artichoke, cabbage, spinach, celery) and strawberries. In 1995, the Monterey County Water Resources Agency (MCWRA) passed an ordinance prohibiting extraction of groundwater between sea level and −76.2 m in Salinas and Castroville. In 1998, Monterey County Water Recycling Projects (MCWRP) began delivering recycled water to 486 hectares (12,000 acres) in the northern Salinas Valley. Crop rotations and management practices at the eight sites of the study area are listed in Table A1; the crops include lettuce, broccoli, cauliflower, cabbage, celery, spinach, artichokes, and strawberries. Drip irrigation was used at the control site and vegetable crops were established with sprinklers for 20 to 30 days. At site 2, sprinkler irrigation switched to drip after plant establishment in 2002 while site 3 used sprinklers for vegetables and drip for strawberry. Sites 4 and 5 used sprinklers and drip and sites 6 and 7 used sprinklers for plant establishment and followed by furrow irrigation.

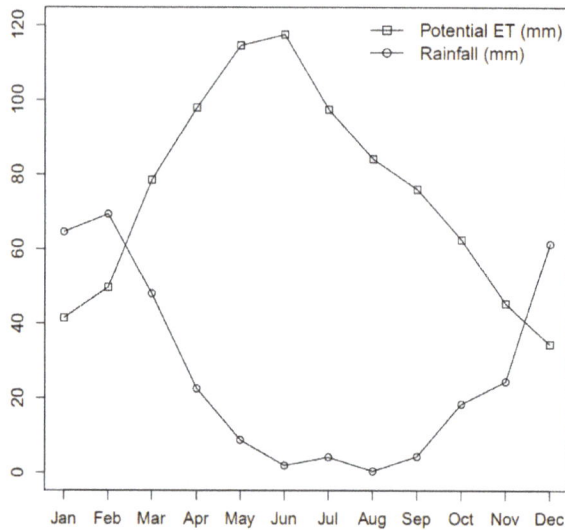

Figure 1. Mean monthly rainfall (P) and reference ET_o from the California Irrigation Management Information System (CIMIS) station 19 in Castroville, CA.

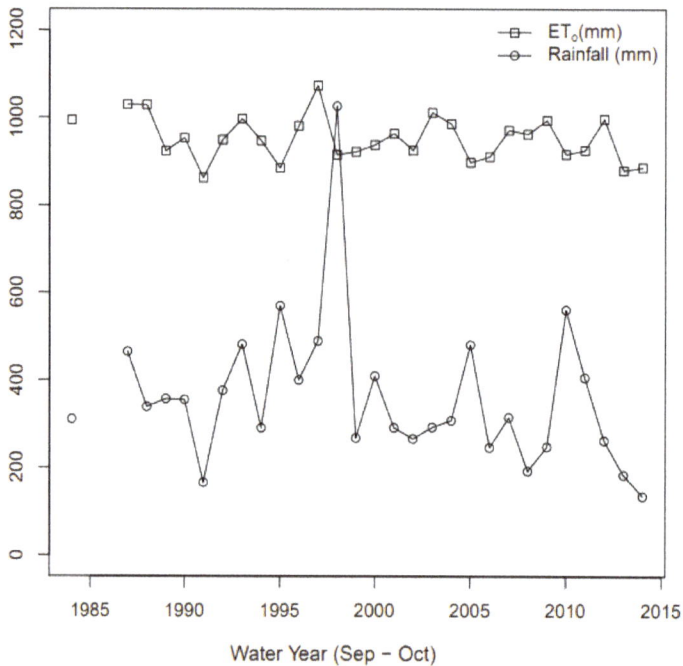

Figure 2. Annual rainfall and reference ET_o from CIMIS station 19 in Castroville, CA.

Tertiary treated wastewater effluent from Monterey Regional Water Pollution Control Agency, (MRWPCA) was sampled on a weekly basis to determine the levels of salt present in it before blending with the supplemental well water used to meet peak irrigation demand. Monthly delivery system sampling confirmed the quality of the water received by growers after supplemental well water was added to the recycled water. In addition, the quality of the well water delivered to the control site was sampled monthly. The water samples were analyzed for pH, EC_w, Na, Mg, Cl and K (potassium) by an accredited laboratory run by MRWPCA. The one control and seven test sites were randomly distributed throughout Castroville, USA area and were chosen based on soil characteristics, drainage systems, types of crops grown (lettuce, cole crops and strawberries), irrigation method and farming practices. At each site, soil samples were collected from depths of 0.03 to 0.30 m, 0.30 to 0.61 m and 0.61 to 0.91 m at four different locations within 1 m of a designated global positioning system (GPS) point. Sample analysis was done by an independent accredited lab (Valley Tech, Tulare, CA, USA) and included pH, soil water electrical conductivity (EC_{sw}), extractable cations B (boron), Ca, Mg, Na and K, and extractable anions Cl, NO_3 (nitrate) and SO_4 (sulphate).

Irrigation water salinity varied between sites and years as recycled water was blended with groundwater (2000–2009) or diverted Salinas River water (2010–2012) and the average applied water EC (EC_w) at the different sites for these time periods are summarized in Table 2. Tertiary treated wastewater in the study had on average Sodium Adsorption Ratio (SAR) value of 5.58, containing 192.1 mg/L of Na, 246.1 mg/L of Cl, and EC of 1.4 dS/m. The rain salinity (EC_p) values were taken from the National Atmospheric Deposition Program station in the Pinnacles National Park located ~322 km east of the study site. The EC_p varies by month ranging from 0.001 dS/m to 0.004 dS/m with May having the highest ion deposits with rain.

Table 2. Average electrical conductivity (EC) of applied water (EC_w) at different sites from 2000–2012.

Site #	2000–2009		2010–2012	
	% Recycled Water	EC_w (dS/m)	% Recycled Water	EC_w (dS/m)
Control	0	0.63	0	0.78
2	46	0.75	92	1.12
3	94	1.52	98	1.19
4	58	0.94	96	1.17
5	93	1.51	100	1.21
6	70	1.14	90	1.09
7	96	1.56	90	1.17

2.2. Root Zone Salinity Model

We coupled crop growth and soil water models applied across the root- and vadose-zones to simulate both upward flow from shallow water tables as well as downward percolation to the groundwater (see Appendix B for a detailed description of the model). These were combined with a root zone salinity model and used to predict root zone soil salinity (see model configuration in Figure 3). The two driving criteria for model selection included the simplification required to describe the processes mathematically without losing the detail needed to develop realistic results, and the model reliance on readily available input data. The Isidoro and Grattan [17] daily time-step model uses a closed-form solution of Reference [32] to describe vertical unsaturated water movement in the root zone and unsaturated zone (Equations (A1) to (A4)). A number of closed-form formulas have been proposed to empirically describe the dependence of unsaturated hydraulic conductivity and water content on pressure head [33–37]. We used the Clapp and Hornberger equation [32] to extend vertical flow through a continuous soil profile to compute the movement of water and salt across the entire vadose zone to the account for shallow groundwater flow processes. Thus, two additional layers from the root zone to the groundwater table were added for both unsaturated and saturated zones.

The crop component of the model includes crop development stages (Table A2), root growth (Equation (A8)), root water uptake and water stress response functions (Equations (A5)–(A7)). Evapotranspiration (ET) includes a combination of two separate processes whereby water is lost from the soil surface by evaporation and from the root zone by crop transpiration. The crop ET is calculated as the product of the reference ET_o and the estimated crop coefficient (K_c) that depends on crop characteristics, vegetative growth state, canopy cover, and height as well as surface-soil properties (Equations (A5)–(A7). In each layer (k), the actual crop ET can be lower than potential ETc(k) due to water stress, which depends on the soil water content and the sensitivity of the crop to low water contents, accounted for through the crop-specific parameter *p*: the ratio of readily available soil water (RAW) to total available water (TAW) (p = RAW/TAW) [38].

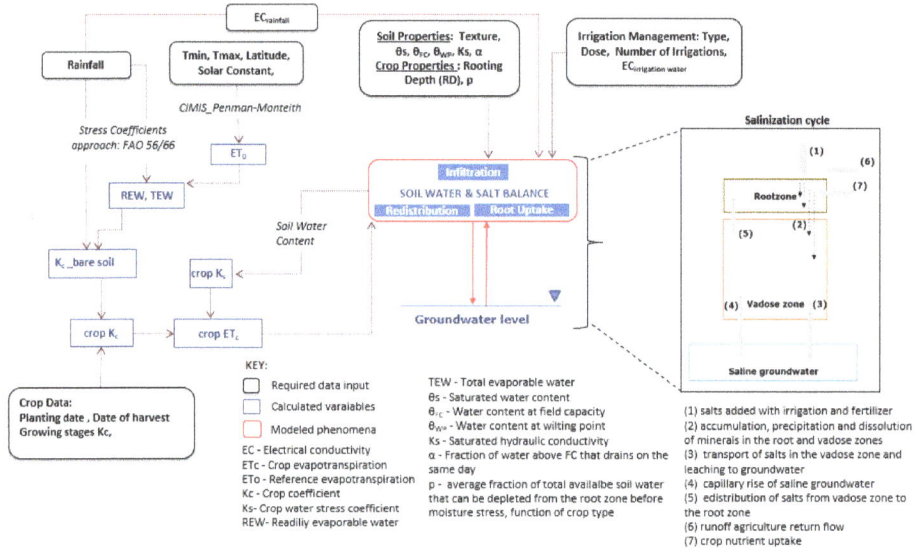

Figure 3. Main components of the soil–water and salinity balance model across the soil–water–plant–atmosphere–aquifer continuum.

The model domain consists of a one-dimensional vertical 7.7-m soil profile, representing the crop root zone and the unsaturated zone that overly a fixed saturated (water table) zone. The domain is discretized so that the clay loam root zone is divided into four quarters of equal depth to enable determination of plant water uptake fractions. The unsaturated zone below the root zone is divided into two layers: a sandy-loam layer immediately below the root zone and a silty clay-loam between the sandy-loam and the capillary fringe. Both upward flow from and downward flow to shallow groundwater is possible in the model. Surface runoff depths were calculated using the Soil Conservation Service (SCS) runoff method [39,40] see Equations (A9) and (A10). Equations (A11) and (A12) detail the soil–water balance simulations.

Salt balance calculations were performed in conjunction with the soil–water balance assuming complete mixing of water entering each layer with that already stored in that layer (Equations (A13)–(A19)). The soil–water EC is used as a salinity indicator, implicitly assuming there is a unique relationship between EC and total dissolved solids (TDS), and that EC behaves as a non-reactive (conservative) solute. Salinity of irrigation water (ECw) and precipitation (ECp) are input values. The mass of salts in layer k (Z(k)) is estimated from the product $EC_{sw}(k) W(k)$, where EC_{sw} is the electrical conductivity of the soil water in that layer.

Plant water uptake was assumed to be a descending extraction pattern that depends on irrigation frequency such that greater uptake is at the top quarter of the root zone [18,41–43]. Plant growth and root development parameters are summarized in Table A2 and Equation (A8) [38,44–47] and we assumed that strawberries were planted on 1.3 m wide raised beds as is common in the region. The model was calibrated for soil salinity generation due to dissolution and a dissolution rate is used to account for these processes. Irrigation and rainwater salinity are specified by the user and the model neglects plant root uptake of salts. While preferential flow and irrigation non-uniformity may also be important features, they were beyond the scope of this model.

Each simulation extended for a period of 13 and 50 water years (1 October to 30 September) and more importantly the model simulates carry over effect from one year to the next. The surface boundary conditions of rainfall, irrigation and ET_o were specified daily together with irrigation and rain water EC. The lower boundary conditions were specified as fixed water table depth and groundwater salinity ECgw; though fixed water table depths are unlikely in the field, water table fluctuations are assumed to be dampened by capillary rise and evaporation from the water table. The model is written in R to make it more widely accessible to water managers and possibly growers.

2.3. Calibration and Sensitivity Analyses

Parameter sensitivity analyses provide insight into those model parameters that are most critical towards approximating measured results and are often used to help focus field sampling or measurement efforts and/or refinement of modeled processes. Here, we take a different approach and first used a global sensitivity analysis that considers variations within the entire variability space of the input factors. We used the elementary effect (EE) method for screening important input factors among the 33 factors initially considered important. Finally, the variance-based "Sobol" method was used with those factors determined to be significant from the Morris screening method for factor fixing and to identify those factors which, left free to vary over their range of uncertainty, make no significant contribution to the variance of the output results of interest. We applied the modified Isidoro and Grattan model to the 13 years of soil salinity observations up to 91.4 cm below the ground surface. Measured data from the control site, irrigated only with available groundwater, was used for calibration and one of the other eight test sites using predominantly recycled water (94–98% recycled/freshwater blend) was used for model validation so as to bracket possible model predictions. The calibrated model was then used to assess the long-term (50 years) salinity outcomes of variable management strategies including cropping patterns, irrigation technology, and irrigation water quality.

We use the Morris and Sobol's methods to support model calibration as shown in Figure 4. The sensitivity analysis was used to address the following questions:

- What input parameters or factors cause the largest variation in the output?
- Are there any factors whose variability has a negligible effect on the output?
- Are there interactions that amplify or dampen the variability induced by individual factors?

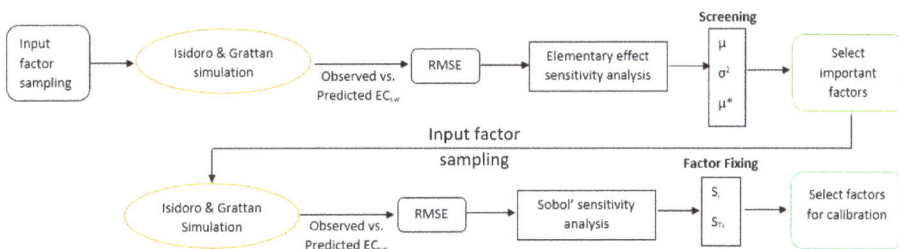

Figure 4. Workflow for calibration of the soil-water water and salt balance model.

Sensitivity Analysis Library (SALib version 1.1.3) an open source library written in Python, was used for performing the sensitivity analyses [48]. The model input variables considered for the sensitivity analysis are listed in Table 3 and parameters showing the greatest sensitivity were selected for calibration.

Table 3. Model parameters used in the sensitivity analysis. (Note: rz—root zone layers 1–4; 5—unsaturated zone layer 5; 6—unsaturated zone layer 6).

Property	Model Code	Model Units	Range Min.	Range Max.
		Soil Hydraulic Parameters		
Saturated water content	Tsrz	cm^3/cm^3	0.439	0.486
	Ts5	cm^3/cm^3	0.357	0.37
	Ts6	cm^3/cm^3	0.36	0.38
Water content at field capacity	Tfcrz	cm^3/cm^3	0.324	0.367
	Tfc5	cm^3/cm^4	0.240	0.320
	Tfc6	cm^3/cm^5	0.250	0.270
Water content at wilting point	Twprz	cm^3/cm^3	0.154	0.177
Residual water content	Tr5	cm^3/cm^5	0.11	0.14
	Tr6	cm^3/cm^5	0.066	0.08
Saturated hydraulic conductivity	Ksrz	cm/day	12.1	36.3
	Ks5	cm/day	36.3	121.8
	Ks6	cm/day	3.6	12.1
Fraction of excess water drained the first day	Arz	%	0.81	0.83
	A5	%	0.85	0.87
	A6	%	0.8	0.82
Runoff Curve Number for fallow periods	CNf	-	91	94
Growing season Curve Number	CNc	-	88	91
Capillary fridge height above the water table	H_d	cm	83	183
Depth to groundwater table	Hwt	M	7.45	11.57
Depth of surface soil layer subjected to drying by evaporation	Ze	mm	100	150
		Plant Parameters		
Root water uptake for layers 1–3	RWU_1	%	60	71
	RWU_2	%	20	30
	RWU_3	%	6	7
Rooting depth of lettuce, broccoli, cabbage and cauliflower	ZrL	cm	30	50
	ZrBrc	cm	40	60
	ZrCabb	cm	50	80
	ZrCau	cm	40	70
Fraction of total available water that can be depleted from the root zone before moisture stress for lettuce	pL	-	0.3	0.7
Fraction of total available water that can be depleted from the root zone before moisture stress for broccoli, cabbage, and cauliflower	pBCC	-	0.45	0.7
		Soil Chemical Parameter		
Rate of dissolution at 2.5–30.5 cm depth	k1	dSm^{-1}/day	0	0.014
Rate of dissolution at 30.5–61 cm depth	k2	dSm^{-1}/day	0	0.022
Salinity at shallow water table	ECgw	dSm^{-1}	0.35	1.58
Initial ECsw	ECsw	dSm^{-1}	0.29	4.1

Different crops have different water uptake patterns, but all take water from wherever it is most readily available within the rooting depth. The root zone water-uptake pattern depends on irrigation frequency. With infrequent irrigations, the typical extraction fractions by root zone layer is 40–30–20–10%. For frequent drip or sprinkler irrigation, the water uptake fractions are skewed towards greater uptake from the upper root zone, or a 60–30–7–3% uptake pattern [18]; this pattern is assumed in many classical analysis of saline soils [42]. Some have suggested use of an exponential model that specifies a greater proportion of uptake near the soil surface, that is, uptake fractions of 71–20–6–3% [41,43]. Ranges for the crop related data, including crop coefficients (Kc), rooting depths (Zr), and average fractions of total available water that can be depleted from the root zone before moisture stress (p), were taken from the Food and Agriculture Organization of the United Nations (FAO)-56 [38]. Estimates of strawberry crop coefficients were found in Reference [47].

All field sites considered were situated on Pacheco clay, clay-loam, and sandy loam soils with ranges of soil texture, available water content, bulk density, organic matter content, and saturated hydraulic conductivity (Ks) taken from SSURGO soil surveys as summarized previously in Table 1. Soil hydraulic properties required for model application were inferred from the soil survey information. We fitted that information to the van Genuchten model using a Multiobjective Retention Curve Estimator (MORE) based on the Multiobjective Shuffle Complex Evolution Metropolis (MOSCEM-AU) algorithm implemented in HYDRUS-1D [30]. The fraction "α" of the excess water that drained the first day ($0 < \alpha < 1$) was calculated from the soil texture in the layer through an empirical relation obtained to match the results presented in [49]. Grismer [20] provides a relationship between capillary fringe heights (Hd) and saturated intrinsic permeability for different soil textures. Groundwater levels were taken from regular measurements by the MCWRA at well 13S/02E-32E05 about 5 km west of study area. Estimated range of groundwater salinity was taken from Reference [50].

The primary output variable of concern was the soil–water EC_{sw} as determined for root zone layers up to 0.91 m deep. As the Isidoro and Grattan model is a dynamic model, the "output" term in the sensitivity analysis does not refer to the range of spatial and temporal distribution of EC_{sw} but to a summary variable. In this case, the root mean square error (RMSE) that is obtained as a scalar function of the simulated time series output EC_{sw} values. As such, for calibration, the objective function minimized the RMSE associated with model prediction.

2.3.1. Model Parameter Screening-Elementary Effects Method

The elementary effects (EE) method is an effective way of screening for important input factors contained in a model [51]. The fundamental concept of this method involves deriving measures of global sensitivity from a set of local derivatives, or elementary effects, sampled on a grid throughout the parameter space [52]. It is based on one-at-a-time (OAT) analysis, in which each parameter X_i is perturbed along a grid of size Δ_i to create a trajectory through the parameter space. For a given value of X, the elementary effect of the ith input factor is defined as:

$$EE_i = \frac{[Y(X_1, X_2, \ldots, X_{i-1} + X_i + \Delta, \ldots, X_k) - Y(X_1, X_2, \ldots, X_k)]}{\Delta} \tag{1}$$

where $Y(X_1, X_2, \ldots, X_k)$ is a prior point in the trajectory and $X = X_1, X_2, \ldots, X_k$ is any selected value in the parameter space such that the transformed point is still in the parameter range for each index $i = 1, \ldots, k$. The sensitivity measures μ and σ are the mean and the standard deviation of the distribution of EE_i proposed by Morris. Mean parameter (μ) assesses the overall influence of the factor on the output parameter of interest; σ assesses the extent to which parameters interact. Thus, a small σ value implies that the effect of X_i is almost independent of the values taken by other factors; on the other hand, a large σ indicates that a factor is interacting with others because its sensitivity changes across the variability space. Campolongo and others [53] suggest that μ^* is a good proxy of the total sensitivity index, a measure of the overall effect of a factor on the output parameter inclusive of interactions. We analyzed μ^* for all input factors to screen out non-influential factors, and then performed a variance-based analysis with the remaining important factors. Once trajectories are sampled, the resulting r elementary effects per input are available, the statistics μ, $\sigma2$ and μ^* for each factor are computed as:

$$\mu_i = \frac{1}{r} \sum_{j=1}^{r} EE_i^j \tag{2}$$

$$\mu_i^* = \frac{1}{r} \sum_{j=1}^{r} |EE_i^j| \tag{3}$$

$$\sigma_i^* = \frac{1}{r-1} \sum_{j=1}^{r} \left(EE_i^j - \mu\right)^2 \tag{4}$$

2.3.2. Factor Fixing-Sobol's Variance Method

Sobol's sensitivity analysis is a global-variance based method. Sensitivity measures are based on the decomposition of the model output variance to individual parameters and the interaction between parameters [54,55]. Variance-based sensitivity analysis relies on three principles:

- input factors are assumed to be stochastic variables of the model that induce a distribution in the output space;
- the variance of the output distribution is a good proxy of its uncertainty; and
- the contribution to the output variance from a given input factor is a measure of sensitivity.

Contribution to total output variance by individual input factors and their interaction can be written using an ANOVA high-dimensional model representation (HDMR) decomposition [51]:

$$V(Y) = \sum_{i}^{k} V_i + \sum_{1 \le i < j \le k} V_{ij} + \ldots + V_{12\ldots k} \tag{5}$$

where $V(Y)$ is the total or unconditional variance of the output, the conditional variance; V_i is the conditional variance or first-order effect of X_i on Y; V_{ij} is the joint effect of X_i and X_j minus the first order-effects for the same factors. Several variance-based indices can be defined; the first order index represents the main contribution effect of each input factor to the output variance can be determined from:

$$S_i = \frac{V_i}{V(Y)} \tag{6}$$

The total order index, S_T, a measure of the overall contribution to output variance from an input factor considering its direct effect and its interactions with all other factors and is determined from:

$$S_{Ti} = 1 - \frac{V_{\sim i}}{V(Y)} \tag{7}$$

where $V_{\sim i}$ is the conditional variance with respect to all the factors but one, i.e., $X_{\sim i}$. The condition $S_{Ti} = 0$ is necessary and sufficient for X_i to be a noninfluential factor on the output. That is, if $S_{Ti} \cong 0$, then X_i can be fixed at any value within it range of uncertainty without appreciably affecting the value of the output variance $V(Y)$. Here, we calculated all of the indices to determine the factors that can be fixed in the calibration process. A recommended sampling technique uses sequences of quasi-random numbers generating n, 2k matrix of random numbers where n is called a base sample and k is the number of input factors. This scheme allows for n (k + 2) model evaluations. We evaluated up to $n = 12$ and found no changes occurred after $n = 10$ and concluded that it was sufficient.

2.4. Long-Term Salinity Indicators

Salt output loads with deep percolation were used to assess the potential for groundwater resources deterioration and salt output loads with runoff indicate the salinity threats posed by treated wastewater irrigation on the salinization of the Salinas River. Similarly, crops respond to the salinity in the root zone over the entire growing season [18]. Thus, we used seasonal-averaged root zone EC of the saturated extract or EC_{eS} (dS/m), deep percolation S_d (kg/ha), and surface runoff S_r (kg/ha) salt output loads as key output state variables to describe long-term (decadal) impacts of irrigation with treated wastewater and farm management practices (i.e., applied water quality and depths, irrigation technology, and crop rotations). Annual rainfall mitigates impacts of irrigating with saline water as such accounting for rainfall leaching is important for evaluating long-term dynamics.

Daily EC of the saturated extract (EC_e) in the root zone layers (k) is calculated as:

$$EC_e(k)_t = \frac{ECsw(k)_t \times \theta(k)_t}{SP(k)} \tag{8}$$

where SP(k) is the saturation percentage (water content of the saturated soil paste expressed on a dry weight basis) for layer k. Traditionally, for most mineral soils it is assumed that field capacity is half of SP, so EC_{et} is the mean EC_e of the 4 rootzone layers; daily values EC_{et} are then averaged over the entire growing season yielding the seasonal-averaged root zone EC_e (EC_{eS}):

$$EC_{eS} = \frac{\Sigma_{\text{Growing season}} EC_{et}}{\text{Days in the growing season}} \tag{9}$$

We applied the model to both meteorological series and management practices for 13 years of cropping practices in the study area. The objective was to simulate a 13-year continuous cropping and provide a multiple-year record of seasonal EC_{eS} and related parameters. Following [15], we assumed a factor of 640 to convert EC (dS/m) into TDS (mg/L) for EC $\leq$ 5 dS/m and a factor of 800 for EC > 5 dS/m.

The model estimates daily lower boundary water flux (D) along with an estimate of soil–water EC in the bottom layer to determine the daily salt output load associated with deep percolation water calculated as:

$$S_d = D \times EC_{sw}(4) \times 6.4 \tag{10}$$

where S_d is the daily drainage salt output load in kg/ha; $EC_{sw}(4)$ is the EC of soil water in the bottom root zone layer in dS/m and 6.4 is the conversion factor assuming a factor of 640 to convert EC (dS/m) into TDS (mg/L) and flux in mm. Additionally, the model estimates daily runoff volumes (SR) from the soil surface along with the runoff water EC such that the daily salt output load associated with runoff is calculated as:

$$S_r = SR \times EC_{sr} \times 6.4 \tag{11}$$

where S_r is the daily runoff salt output load in kg/ha; EC_{sr} is the EC of surface runoff in dS/m and 6.4 is the conversion factor.

2.5. Calibration and Validation

The primary objective of model calibration was to capture the long-term soil salinity dynamics in the fields irrigated with treated wastewater in the study region by satisfactorily reproducing the 2000 to 2012 soil-water salinity (EC_{sw}) data set described by [11]. The study consisted of six test sites and one control site randomly distributed across the Castroville region that were chosen to provide a typical range of soil characteristics, drainage systems, types of crops grown, irrigation methods, and farming practices found in the region. Average annual water quality delivered to each site was determined as well as soil samples collected from depths of 0.03–0.30 m, 0.30–0.61 m, and 0.61 to 0.91 m at four different locations within 1 m of a designated global positioning system (GPS) point.

Soil samples were collected following winter rains before the spring planting, during and at the end of the summer growing season prior to winter rains. Saturated paste extracts from these soil samples were analyzed for EC and solute concentrations. The control site received only well (2000–2009) or surface (2010–2012) water and this site was used for calibration, while site 3 received 94–98% recycled water and this site was used for validation (see Table 2). Annual crop rotations that included lettuce, broccoli, cauliflower, cabbage, and strawberry are shown in Figure A1. Control site vegetables were established with sprinklers for 20–30 days and drip irrigated, while at site 3, vegetables were sprinkler irrigated and then drip irrigation was used for strawberries.

Sensitivity analysis, calibration and validation were performed for soil–water EC (EC_{sw}) at three depth intervals. The goodness-of-fit measures of the model predictions were evaluated using a set of statistical indices including root-mean-squared-error (RMSE), Nash-Sutcliffe efficiency (NSE), coefficient of determination (R^2), coefficient of regression (b), and mean relative error (MRE). A perfect fit between observations and model predictions yields a RMSE = 0.0, NSE = 1.0, R^2 = 1.0, b = 1.0 and MRE = 0.0.

3. Results

3.1. Sensitivity Analysis

The sensitivity analyses were performed on the 33 model parameters as indicated in Table 3. We measure the sensitivity of the root mean squared error (RMSE) metric, calculated using the sampled soil water salinity (EC_{sw}) at the three depth intervals to ensure that our sensitivity indices are grounded relative to the observed soil salinity. The sample sizes and corresponding number of model evaluations required for both the Elementary Effect (EE) and Sobol' methods are listed in Table 4. For the EE method, a sample size of n = 1000 with 10 optimal trajectories were used resulting in 340 model evaluations. Since the Sobol analysis was performed with the reduced parameter space results determined from the prior EE approach, the number of samples selected ranged from 10 to 12 and iterations were discontinued at the point where the sensitivity results converged. The Latin Hypercube sampling design was used with n = 50 sampling points to generate a near-random sample of parameter values as proposed by Reference [56]. The open-source implementations were used for both methods [48]. Convergence was considered acceptable when within the 95% confidence interval.

Table 4. Sampling sizes and number of model runs performed for each sensitivity analysis method.

Method	Sample Size	Model Evaluations
Elementary Effect (EE) w/10 trajectories	1000	340
Sobol's	10	200
	11	220
	12	240

Among the 33 rootzone model parameters (Table 3), the EE method revealed that only nine strongly influenced the output soil–water EC (EC_{sw}) as summarized in Table 5. Table 5 indicates parameter sensitivity values μ, μ^*, μ^*_conf and Σ/EE from the elementary effect analysis. The mean of the distribution of EEi (Σ/EE) is a proxy for a total sensitivity index of ith input factor and μ^*_conf is the confidence interval of the μ^* at 95%. We chose these nine parameters based on the combination of small μ^* values with corresponding large σ values (Figure 5). High σ value indicates that the EE are strongly affected by the choice of the sample points at which they are computed, therefore a non-negligible interaction with other factor values as illustrated in Figures 5 and 6.

Next, we applied Sobol's sensitivity analysis to determine factor fixing using the nine influential factors from the EE results as summarized in Table 6, we reported the first-order (Si) and total order sensitivity (S_{Ti}). Of the nine factors, only two were non-influential by this analysis, that is, they resulted in $S_T = 0$. These non-influential parameters towards average root zone salinity included the capillary fringe height (Hd) and depth to groundwater (Hwt), so these were fixed at average values for the model calibration analysis. Interestingly, S_T values for Tsrz for layer 2 and layer 3 indicate that saturated soil moisture content is more influential for these layers than Ks, and especially that Tsrz in layer 2 is the most influential parameter in the model. Given that we used the control site for model calibration and irrigation of vegetables for this site is managed with sprinklers for establishment and then drip for the rest of the plant development stages, we suspect that plant water uptake is mainly from the bottom layers. However, in our model plant water uptake even with drip is assumed to be mainly from the top layer: 60–30–7–3% uptake pattern. Root water uptake (RWU) although included in the sensitivity analysis was not influential. Mostly likely, upward flow of soil water from the second layer provides water required for uptake in the top layer. Soil moisture is directly related to unsaturated conductivity (a driving force for soil water from wet to dry soil layer). As such the water saturated water content in the second layer ended up being most influential in our calibration. It is important to note that with respect to the crop rooting depth, it is likely that the lettuce rootzone depth is important for the control site specifically as it was the main crop grown for the majority of the experiment period. For other sites, it may be important to include variability of the crop rooting depth.

Table 5. Parameter sensitivity based on Morris indices.

Code	Parameter	μ*	μ	μ*_Conf	Σ/EE
	Root zone Sampling Depth of 0.03–0.3 m				
k1	Rate of dissolution at 0.03–0.30 m depth	0.16	−0.14	0.07	0.15
ZrL	Rooting depth of lettuce	0.1	−0.10	0.06	0.1
Tsrz	Root zone saturated water content	0.09	−0.07	0.08	0.14
Ksrz	Root zone saturated hydraulic conductivity	0.08	0.03	0.06	0.12
Twprz	Root zone water content at wilting point	0.05	−0.05	0.03	0.05
Hd	Capillary fringe height	0.04	0.04	0.08	0.12
pL	Fraction of depletable moisture for lettuce	0.04	0.03	0.02	0.04
	Root zone Sampling Depth of 0.30–0.61 m				
k1	Rate of dissolution at 0.03–0.30 m depth	0.12	−0.1	0.07	0.13
Ksrz	Root zone saturated hydraulic conductivity	0.07	−0.03	0.04	0.11
Tsrz	Root zone saturated water content	0.07	−0.05	0.06	0.12
Tfc5	Unsaturated zone saturated water content	0.05	−0.02	0.04	0.08
k2	Rate of dissolution at 0.3–0.61 m depth	0.05	−0.05	0.02	0.04
ZrL	Rooting depth of lettuce	0.05	−0.04	0.03	0.06
Twprz	Root zone water content at wilting point	0.04	−0.02	0.02	0.05
Hd	Capillary fringe height	0.04	0.03	0.06	0.1
Tr5	Unsaturated zone residual water content	0.03	−0.03	0.02	0.04
	Root zone Sampling Depth of 0.61–0.91 m				
k1	Rate of dissolution at 0.03–0.30 m depth	0.13	−0.08	0.06	0.16
Tsrz	Root zone saturated water content	0.09	−0.05	0.06	0.14
Ksrz	Root zone saturated hydraulic conductivity	0.08	−0.01	0.06	0.13
Tfc5	Unsaturated zone saturated water content	0.06	−0.03	0.06	0.11
Twprz	Root zone water content at wilting point	0.05	−0.03	0.03	0.07
Hwt	Depth to groundwater table	0.05	0.05	0.06	0.1
ZrL	Rooting depth of lettuce	0.05	−0.04	0.04	0.07
k2	Rate of dissolution at 0.3–0.61 m depth	0.05	−0.04	0.03	0.04
Hd	Capillary fringe height	0.04	0.04	0.07	0.13
Tr5	Unsaturated zone residual water content	0.03	−0.03	0.02	0.04

μ* is a good proxy of the total sensitivity index [53].

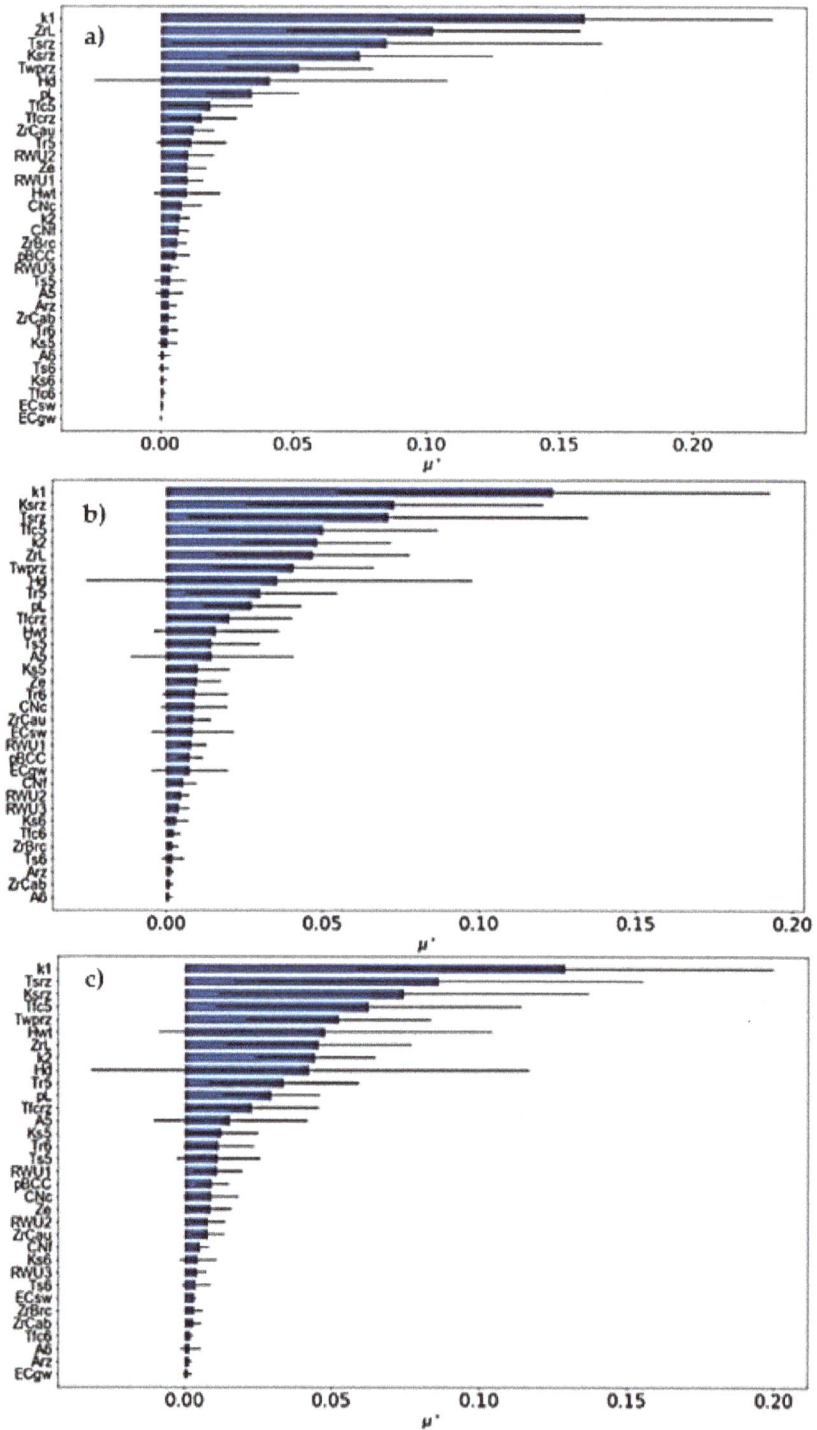

Figure 5. Morris method μ* (the absolute of the mean elementary effect) values for soil depth layers. (**a**) 0.03–0.30 m; (**b**) 0.3–61 m; (**c**) 0.61–0.91 m depths.

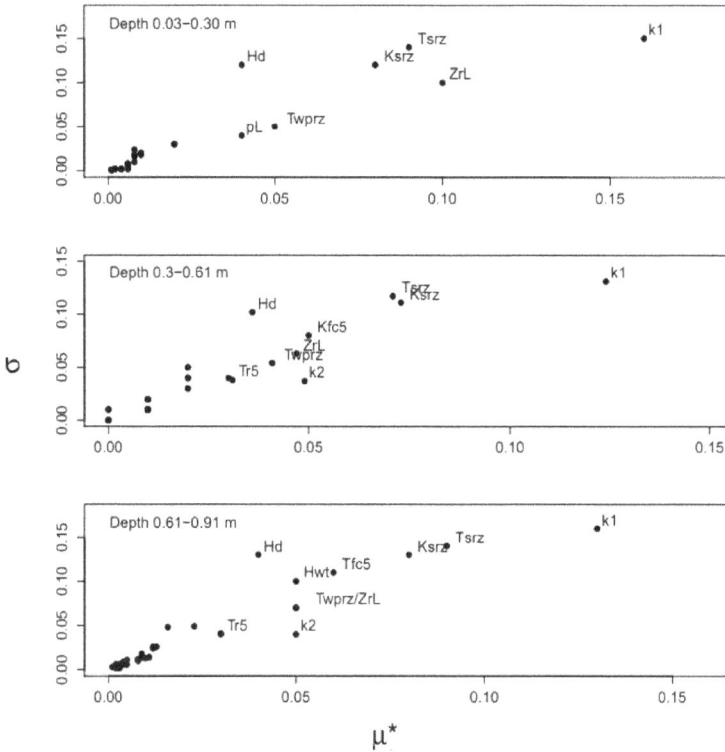

Figure 6. Morris method μ* (the absolute of the mean elementary effect) vs. σ (the standard deviation of the elementary effect) for the different soil depths.

Table 6. Sobol's sensitivity indices.

Code	Parameter	S_T	S_T_conf	Si	Si_conf
	Root zone Sampling Depth of 0.03–0.3 m				
k1	Rate of dissolution at 0.03–0.30 m depth	0.65	0.42	0.42	0.33
ZrL	Rooting depth of lettuce	0.43	0.47	−0.10	0.29
Tfcrz	Unsaturated zone saturated water content	0.07	0.06	−0.04	0.13
Ksrz	Root zone saturated hydraulic conductivity	0.05	0.08	−0.01	0.06
Tsrz	Root zone saturated water content	0.04	0.05	−0.07	0.11
Twprz	Root zone water content at wilting point	0.02	0.02	−0.08	0.07
k2	Rate of dissolution at 0.3–0.61 m depth	0.01	0.01	0.03	0.05
Hwt	Depth to groundwater table	0	0	0	0.02
H$_d$	Capillary fringe height	0	0	0	0.02
	Root zone Sampling Depth of 0.30–0.61 m				
Tsrz	Root zone saturated water content	3.38	9.1	0.01	0.11
k1	Rate of dissolution at 0.03–0.30 m depth	0.86	0.53	0.31	0.5
k2	Rate of dissolution at 0.3–0.61 m depth	0.27	0.19	0.28	0.28
ZrL	Rooting depth of lettuce	0.24	0.49	0.03	0.24
Tfcrz	Unsaturated zone saturated water content	0.05	0.06	0	0.14
Ksrz	Root zone saturated hydraulic conductivity	0.04	0.07	0.01	0.13
Twprz	Root zone water content at wilting point	0.02	0.02	−0.03	0.07
Hwt	Depth to groundwater table	0	0.01	−0.02	0.05
H$_d$	Capillary fringe height	0	0.01	−0.02	0.05

Table 6. *Cont.*

Code	Parameter	S_T	S_T_conf	Si	Si_conf
	Root zone Sampling Depth of 0.61–0.91 m				
k1	Rate of dissolution at 0.03–0.30 m depth	0.89	0.68	0.05	0.44
ZrL	Rooting depth of lettuce	0.36	0.48	0.05	0.21
k2	Rate of dissolution at 0.3–0.61 m depth	0.29	0.17	0.09	0.24
Tsrz	Root zone saturated water content	0.12	0.22	0.02	0.08
Ksrz	Root zone saturated hydraulic conductivity	0.07	0.1	−0.03	0.08
Tfcrz	Unsaturated zone saturated water content	0.06	0.07	0.03	0.1
Twprz	Root zone water content at wilting point	0.03	0.03	−0.01	0.04
H_d	Capillary fringe height	0	0.01	0	0.01
Hwt	Depth to groundwater table	0	0.01	0	0.01

3.2. Calibration and Validation

Model calibration was performed allowing the parameters k1, k2, ZrL, Tsrz, Ksrz, Tfcrz, and Twprz identified as most sensitive to model determination of soil–water EC (EC_{sw}) to vary within their ranges and output compared to the measured soil salinity at the control site. Model validation was completed by simulating EC_{sw} for site 3. The inclusion of the saturated soil water content and saturated hydraulic conductivity in the calibration was crucial as infiltration rates were expected to vary with changes in exchangeable sodium in the soil. O'Geen [57] provides classification of salt-affected soils based on trends of soil water EC, exchangeable sodium percentage (ESP), and SAR. We assessed the potential infiltration problems caused by irrigation water quality following Reference [6] (p. 44) and found that slight-to-moderate reduction in infiltration rates due to irrigation water salinity were expected for the control site and site 4 (Figure A2). It is however interesting to note that blending well water and recycled water alleviated the possible adverse effects of well water on soil infiltration rates.

The latin hypercube sampling design was used with $N = 50$ sampling points as proposed by Reference [56]. Intervals were sampled without replacement to ensure even distribution of points with respect to each variable. We executed the model 50 times and computed the corresponding RMSE associated with model predictions. An open-source global optimization code DEoptim written in R was used to find a global minimum RMSE [58]. DEoptim implements the differential evolution algorithm for global optimization. The estimated best fits with the least RMSE values are listed in Table 7.

Table 7. Best-fit parameter values estimated with calibration for EC_{sw}.

Property	Code	Units	Value
Soil Hydraulic Parameters			
Root zone saturated water content	Tsrz	m^3/m^3	0.467
Root zone water content at field capacity	Tfcrz	m^3/m^3	0.361
Root zone water content at wilting point	Twprz	m^3/m^3	0.172
Root zone saturated hydraulic conductivity	Ksrz	mm/day	347
Plant Parameters			
Rooting depth of lettuce	ZrL	m	0.5
Soil–Water Chemistry Parameters			
Rate of dissolution at 0.03–0.30 m	k1	dSm^{-1}/day	0.014
Rate of dissolution at 0.30–0.61 m	k2	dSm^{-1}/day	0.022

Summarized in Table 8 are the indices associated with comparisons between EC_{sw} measured in the field and model predicted values at different soil depths. For all depth intervals, the sum of first-order effects and sum of total order indices is greater than one, indicating that there are interactions among model factors. Moreover, for factors that have total indices greater than their first-order values, other factors are taking part in the interaction such that throughout the soil profile the root zone hydraulic parameters, rooting depth and dissolution rate are taking part in determination of the soil-water EC, with the dissolution rate accounting for the largest fraction of output variance. This observation provides insight about how well water flow and salinity is modeled.

Table 8. Statistical indices for simulated vs. observed soil water EC (EC_{sw}).

Index	RMSE	MRE	NSE	R^2	b
Optimal value	0	0	1	1	1
Control Site (Calibration)					
0.03–0.30 m	0.45	0.23	0.72	0.73	0.89
0.30–0.61 m	0.41	0.25	0.52	0.52	0.89
0.61–0.91 m	0.34	0.26	0.16	0.27	0.95
Site 3 (Validation)					
0.03–0.30 m	0.95	0.24	0.56	0.60	0.87
0.30–0.61 m	0.70	0.21	0.24	0.39	0.88
0.61–0.91 m	0.60	0.20	0.48	0.51	0.94

Model realizations for the calibration and validation runs were compared with measured EC_{sw} and shown in Figure 7. Model predictions did not capture large values of observed EC_{sw} though model performance improved with increased soil depth. The RMSE index indicated large discrepancies between predicted and measured values, hence the greater values at site 3. Whereas R^2 values reflect the combined dispersion against the single dispersion of the observed and predicted values. The mean relative error (MRE < 30%) for all layers indicates satisfactory model performance, while the larger NSE values for the top soil layer indicate that the modeling effort is worthwhile in predicting near surface salinity to depths of 0.3 m.

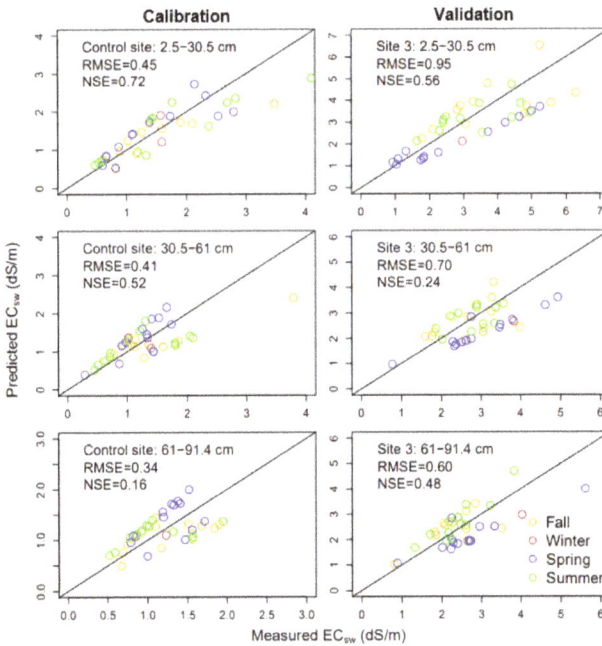

Figure 7. Predicted vs. measured EC_{sw} at 2.5–30.5 cm, 30.5–61 cm and 61–91.4 cm soil depths for control site (calibration) and site 3 (validation).

For all calibrated model runs, regressions of the predicted vs. observed values resulted in a non-zero intercept, b of nearly 1 dS/m. Taken together with the low R^2-values, we conclude that the model persistently underestimates EC_{sw}, especially those observed EC_{sw} values greater than ~2 dS/m. Based on the relatively small MRE values that provide an indication of the magnitude of the error relative to observed values without considering the error direction, the model captures salinity dynamics for all layers in the root zone. However, the Figure 8 plot of residuals vs. predicted EC_{sw} for the calibration and validation results exhibits heteroscedasticity, that is, residuals grow as the predicted EC_{sw} values increase. Overall, this latter observation suggests that although the NSE, RMSE, and MRE statistics show that the model has some predictive capacity, it does not capture some processes apparently involved in the soil salinity dynamics.

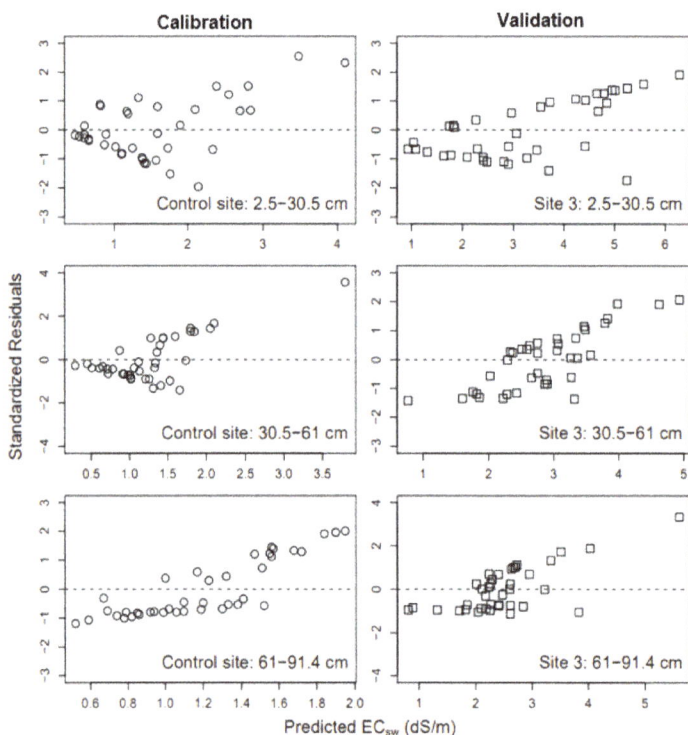

Figure 8. Predicted EC_{sw} vs. standardized residuals at 2.5–30.5 cm, 30.5–61 cm, and 61–91.4 cm soil depths for control site (calibration) and site 3 (validation).

The possible explanation for the differences in measured and predicted soil salinity is that the model does not account for fertilizer and soil amendment management, or plant root uptake of solutes and fertilizer. Generally, transformation (e.g., dissolution) in the soil of different chemicals added during fertigation will increase soil salinity. For example, urea is converted to ammonium that is adsorbed in the soil depending on sol temperature; ammonium is converted to nitrate by nitrification that depends on soil temperature, soil moisture, pH and oxygen content, and nitrate is highly mobile but ammonium, potassium, and phosphorus remain relatively immobile in the root zone. Although the salt index (SI) based on equivalent units of sodium nitrate (developed in 1943 to evaluate the salt hazard of fertilizers) alone cannot be used to evaluate the effect of increased soil salinity from fertilizer applications, it can be used as indicator for the long-term effects on soil salinity. The most commonly used fertilizers in the study region (from California Department of Food and Agriculture annual reports) were nitrogen fertilizers that included urea ammonium nitrate solution (SI = 95), ammonium nitrate (SI =102) and calcium nitrate (SI = 53); phosphorus fertilizer (SI = 7.8-29), potassium sulfate (SI = 46), gypsum, and lime. Sodium nitrate was arbitrarily set at 100, where EC of 0.5 to 40 mass percentage of sodium nitrate is 0.54 to 17.8 dS/m and for a mixture of materials it is reasonable to assume EC is additive for horticulture. As such, the lower the index value the smaller the contribution the fertilizer makes to the level of soluble salts. Thus, fertilizer applications likely add to soil salinities exacerbating the problem over time.

Although the model underestimates EC_{sw}, it adequately captures salinity trends in the leaching of salt during winter months and an increase in salinity water applications and ET during the crop season. In an effort to evaluate performance of transient vs. steady state models, Reference [12] concluded that the transient models better predict the dynamics of the chemical–physical–biological interactions in an agricultural system. However, since we account for irrigation water salinity, rainfall salinity and dissolution of salts in the soil and exclude additions of fertilizer and soil amendment our simulated EC_{sw} values can be viewed as a likely lower bound of soil salinity associated with the irrigation and farm management practices considered in the model description.

Another complexity possible affecting model prediction is the spatial distribution of salinity with drip irrigation as noted by Reference [59]. They used the transient Hydrus-2D model to compare results between field experiments having both drip and sprinkler irrigated processing tomatoes under shallow water table conditions for a wide range of irrigation water salinities. Both field and model results showed that soil-wetting patterns occurring under drip irrigation caused localized leaching which was concentrated near the drip line. In addition, a high-salinity soil volume was found near the soil surface that increased with increasing applied water EC. Overall, localized leaching occurred near the drip line while soil salinity increased with increasing distance from the emitter and with increasing soil depth. Such localized non-uniformities in leaching are not captured in the one-dimensional model but may have affected field soil sampling in the drip–irrigated fields of our study region. That is, soil samples collected some distance from where dripline emitters were previously operating would likely have greater salinities than would otherwise occur under the uniform leaching and dissolution conditions assumed in the model. Nonetheless, it is important to note that this model is user-friendly and less data intensive and it can be very useful for setting reference benchmarks of long-term salinity impacts of using saline water for irrigation.

3.3. Long-Term Salinity with Treated Wastewater Irrigation

We used the calibrated and validated model to simulate the long-term (50-year periods) soil salinity in the fields irrigated with varying fractions of treated wastewater, that is, we applied the model to control site and sites 2 to 7 (Tables 2 and 3). Fifty-year simulations assumed randomly selected rainfall and ET_o data from historical records (1983 to 2014) from 2013 to 2049 and the 13-year cropping patterns and irrigation management. In the model calibration, the groundwater table height (Hwt) and groundwater salinity (ECgw) were found to be non-influential parameters with respect to the measured soil water EC (EC_{sw}). As such average values of measured Hwt and ECgw from the monitoring well located ~5 km west of the study site were used, that is, 0.95 m and 0.97 dS/m, respectively. For each simulated case, the three output variables of interest were averaged root zone salinity over the growing season expressed as EC_{es} (dS/m), annual drainage salt output load as S_d (kg/ha/year), and annual runoff salt output load S_r (kg/ha/year).

Values for the annual average water-balance terms over the 13-water year simulation at each site are summarized in Table 9a and for the 50-water year simulation in Table 9b. Somewhat greater actual crop water uptake (ET) is achieved under drip and sprinkler irrigated fields with similar crop rotations (sites 2 and 3 as compared to 6 and 7). Leaching fractions were generally very low for all sites at less than 2%. The greatest leaching occurred in the vegetable–strawberry rotation that was first sprinkler than drip irrigated, while the other irrigation and cropping practices yielded similar leaching volumes with the exception of no leaching from drip-irrigated artichokes. Similarly, surface runoff is smaller from drip or sprinkler irrigated fields as compared to sprinkler-furrow irrigated fields.

Table 9. Summary of annual average water balance and salinity variables for the different field management scenarios.

				a. 13 Years Simulations				
Site No.	Control Site	2	3	4	5	6	7
Crop management	Vegetables	Vegetables	Vegetables and strawberry	Perennial artichoke		Vegetables and strawberry	Vegetables
Irrigation management	Sprinkler then drip	Sprinkler or drip	Sprinkler then drip	Sprinkler or drip		Sprinkler then furrow	Sprinkler then furrow
Irrigation (mm/yr)	652	597	682	648		694	656
Seasonal Evapotranspiration ET (mm/year)	254	200	252	392		204	177
Surface runoff (mm/year)	299	232	236	158		315	265
Leaching (mm/yr)	7.1	4.4	16.3	0		23.7	12.7
Irrigation water electrical conductivity EC_W (dS/m)	0.71	0.94	1.36	1.06	1.36	1.1	1.37
Root zone EC_{SW} (dS/m)	1.48	1.74	2.45	2.4	2.89	1.76	1.95
Root zone seasonal-averaged electrical conductivity of the saturated extract EC_{eS} (dS/m)	0.09	0.13	0.19	0.15	0.18	0.15	0.17
Salt output load with deep percolation S_d (kg/ha/year)	61	42	207	0	0	153	104
Salt output load with runoff S_r (kg/ha/year)	304	851	1860	648	1012	1073	1465
				b. 50 Years Simulations				
Site No.	Control Site	2	3	4	5	6	7
Seasonal ET (mm/year)	187	128	188	395		141	131
Surface runoff (mm/year)	323	295	236	163		275	250
Leaching (mm/year)	8.0	7.2	9.4	1.0		13.0	8.8
EC_W (dS/m)	0.71	0.94	1.36	1.06	1.36	1.1	1.37
Root zone EC_{SW} (dS/m)	1.12	1.20	1.86	2.05	2.52	1.34	1.50
Root zone EC_{eS} (dS/m)	0.14	0.19	0.27	0.20	0.23	0.24	0.26
S_d (kg/ha/year)	27	20	64	0.8	1.0	56	38
S_r (kg/ha/year)	305	491	1028	339	518	598	764

Simulated average (50-year) annual root zone soil water salinity for all sites is shown in Figure 9. Sites managed with sprinkler or drip irrigation and with higher salinity water (EC_W) had higher estimated annual average EC_{sw}, for example, at sites 3, 4, and 5, EC_{sw} was 2.94 dS/m, 3.15 dS/m, and 3.44 dS/m, respectively. For sites 6 and 7 that used sprinklers for germination then furrow irrigation, the latter site received water with higher salinity than the former 1.37 dS/m compared to 1.1 dS/m on vegetable and strawberry rotation, but the resulting average annual soil water salinity differed little, that is, 2.05 versus 2.89 dS/m, respectively. The control site irrigated with well water had the least annual average rootzone soil–water salinity. Furthermore, soil–water salinity equilibrium $EC_{sw} \leq 2.0$ dS/m was reached throughout the 50-year horizon for the control site irrigated with well water and after 12 years of irrigated with blended wastewater for sites 2, 6, and 7, whereas for sites 3, 4, and 5 soil water EC increased above 2.0 dS/m in the simulation period. Using Mann–Kendall analysis we found that actual ET had a positive and significant association whereas irrigation amounts had a negative and significant association with $EC_{sw} \leq 2.0$ dS/m (Tau = 0.321, *p*-value = 0.016 and

Tau = −0.268, *p*-value = 0.046 respectively). The Mann–Kendall Tau values indicate the strength and direction of monotonic trends, with −1 and 1 representing perfectly negative and positive monotonic trends, respectively, while the *p*-value indicates relative significance [60].

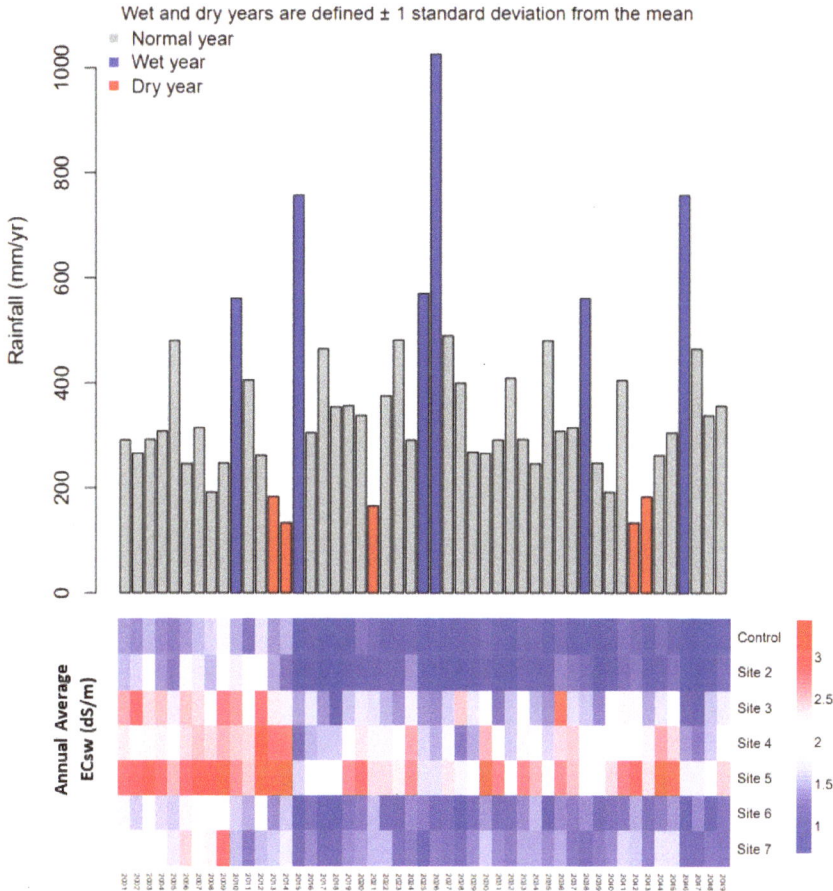

Figure 9. Annual average soil–water salinity for sites irrigated with varying fractions of treated wastewater (2000 to 2049).

As Platts and Grismer [11] found that salt leaching to deeper soil layers occurred during the rainy season (October–March), while during the growing season soil–water EC increases in soil layers near the surface due to evapo-concentration, on the other hand, applied water salinity causes soil–water EC spikes during the growing season. Rainfall was important towards salt leaching from the root zone at all sites as evident during the wet years 2010, 2016, 2025, 2026, 2038, and 2046 when average annual soil water EC decreased. With the exception of sites 3, 4, and 5, there were no upward trends in soil water salinity over the 50-year period. Overall, relatively constant soil–water EC after 50 years simulation of EC_{sw} < initial EC_{sw} of 2.19 dS/m for all sites except site 5 suggest that there was adequate soil leaching in the region for sustained use of the treated wastewater for irrigation. However, the question remained as to what level of soil salinity would be acceptable especially for annual strawberry production.

Crops are generally assumed to respond to seasonal-averaged root zone salinity of the saturated paste (EC_{eS}) and yield loss thresholds and rates of decline with increasing salinity have been determined based on salinity thresholds in [61]. We calculated EC_{eS} for all the sites and plot the range of EC_{eS} in Figure 10. The maximum seasonal-averaged saturated paste EC for each site was 0.19, 0.27, 0.20, 0.23, 0.24, and 0.26 dS/m for sites 2, 3, 4, 5, 6, and 7, respectively. These values are less than half that of the lowest Mass–Hoffman threshold value of $EC_e^* = 1.0$ dS/m associated the most sensitive crop (strawberry) in the rotations considered. As such, it is unlikely that long-term irrigation with treated wastewater in the region will adversely affect crop yields significantly.

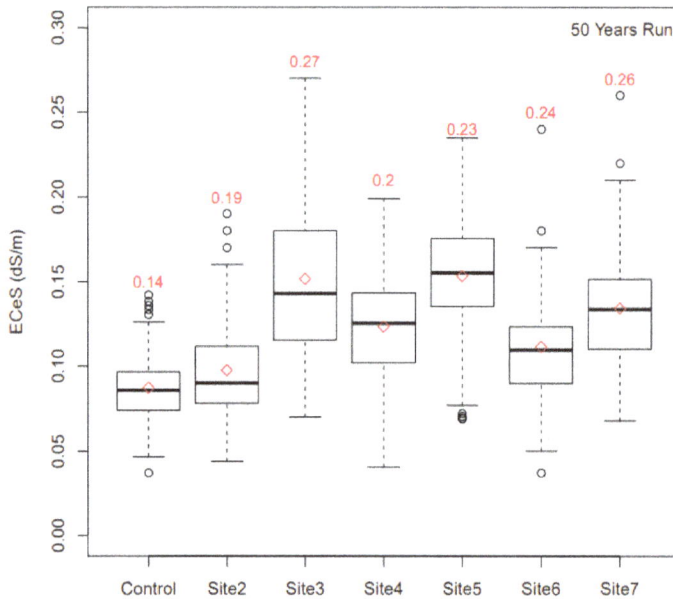

Figure 10. Range of estimated growing season saturated paste EC in the root zone for each site.

In terms of possible adverse environmental effects associated with salinization of surface and ground waters in the region, we determined the cumulative salt output load with deep percolation (S_d) and salt output load with runoff (S_r) during the 50-water year simulation period for the different sites as shown in Figure 11. Salts accompanying surface runoff pose a larger threat in the watershed as these are an order-of-magnitude greater than the cumulative salt output loads to groundwater. Salt loading with deep percolation for the 50-year simulation range up to 3,377 kg/ha with the greatest loads from site 3 and minimal loading for sites 4 and 5 (40 kg/ha and 49 kg/ha respectively) on which annual artichoke crops were grown. Cumulative salt loads accompanying runoff ranged from 19,918 to 59,552 kg/ha, with the greatest loading emanating from site 3 and least from site 4. In comparison to the control site irrigated with well water cumulative salt output loading with deep percolation was 1325 kg/ha and salt output load with runoff 21,505 kg/ha.

To clarify what factors were key to affecting adverse environmental salinization within the region, we tested the effect of applied water EC (EC_w) and depths, rainfall depths, actual crop ET and potential crop ET, and number of days fallow on soil water EC (EC_{sw}) and salt output loads with runoff (S_r) or deep percolation (S_d) using the non-parametric Mann–Kendall trend analysis (the 'Kendall' test in the R package [60]. With respect to the Mann–Kendall Tau values (Table 10), the applied water EC (EC_w) and rainfall depths had positive and significant effects on annual average soil water EC (EC_{sw}) and annual salt output loads with runoff (S_r) and deep percolation (S_d). Calculated actual crop ET had a positive and significant effect on EC_{sw}; this is expected as water uptake by the plant and evaporation leave salts behind. On the other hand, actual crop ET had a negative and significant effect on S_d, and had no significant effect on S_r. Applied water depths had positive and significant effect on S_d and S_r. The number of days fields were fallowed had a negative effect on EC_{sw} and a positive effect on S_d and S_r. Overall, these observations were consistent with that expected from the field observations and described above.

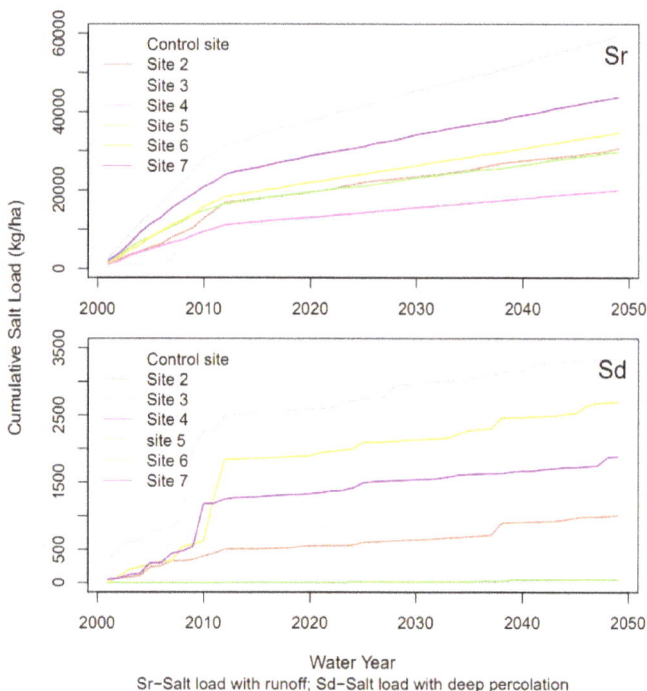

Figure 11. Cumulative salt output loads with runoff and deep percolation from each site.

Table 10. Mann–Kendall trend analysis for average annual soil water EC (EC_{sw}), annual salt output load with runoff (S_r), and with deep percolation (S_d).

Parameter	EC_{sw} (dS/m)		S_r (kg/ha/year)		S_d (kg/ha/year)	
	Tau	*p*-Value [a]	Tau	*p*-Value [a]	Tau	*p*-Value [a]
EC_w (dS/m)	0.33	***	0.49	***	0.04	ns
Rainfall (mm/year)	−0.18	*	0.03	ns	0.06	ns
Potential crop ET (mm/year)	0.03	ns	0.24	**	0.17	*
Actual crop ET (mm/year)	0.44	***	−0.11	ns	−0.33	***
Irrigation (mm/year)	0	ns	0.41	***	0.24	**
Days fallow (days)	−0.44	***	0.26	**	0.39	***

[a] Two-sided *p*-value ranges: $0 \leq$ *** ≤ 0.001; $0.001 <$ ** ≤ 0.01; $0.01 <$ * ≤ 0.05; ns > 0.05.

4. Discussion

We calibrated and validated a modified root zone salinity model originally developed by Isidoro and Grattan [17], which was then applied to estimate long-term soil salinity in fields irrigated with treated wastewater. We conducted a global sensitivity analysis using the elementary effect/Morris and Sobol's methods to first reduce the number of influential model parameters important to calibration and that need to be acquired from the field. Seven of the thirty-three model parameters were found to be critical to root zone soil salinity dynamics. These were parameters accounting for salt dissolution in the soil, root zone hydraulic parameters, and crop rooting depth. Model calibration resulted in a satisfactory fit to the observed field data; however, the model underestimated soil water salinity (EC_{sw}), especially for large $EC_{sw} > 2$ dS/m measured during the growing season. We attributed this error to the model's failure to account for fertilizer and soil amendment applications and transformation thereof (e.g., gypsum dissolution) in the soil. In addition, drip irrigation leads to very localized variations in soil salinity that depend on the distance from the emitter that are not considered in the model and may have affected field soil sampling between plantings and harvests. Nonetheless, the model adequately captured soil–water EC trends that were congruent with observed data.

Sites irrigated with greater salinity water (EC_w) combined with sprinkler or drip had greater estimated annual average soil–water salinity (EC_{sw}). Sites that combined sprinkler irrigation for germination with furrow for the remaining development stages resulted in lower annual average EC_{sw} in the root zone even when more saline irrigation water was applied. Rainfall played an important role in the leaching of salts from the root zone as during wet years average annual soil water EC decreased at all sites. Moreover, rainfall had a negative and significant effect on annual average root zone EC_{sw}. We found that for all sites use of treated wastewater for irrigation over the 50-year period does not affect strawberry yields, the most salt sensitive of the crops in the rotations encountered. Overall irrigation water EC (EC_w), rainfall amounts, actual calculated crop ET and the number of days fields were fallowed had significant effects on annual average soil water EC (EC_{sw}). EC_w, rainfall and actual crop ET effects were positive and fallowing decreased average root zone EC_{sw}. On the other hand, irrigation amounts and number of days fallowed had positive and significant effects on salt output loads associated with runoff and deep percolation. Moreover, soil water salinity equilibrium $EC_{sw} \leq 2.0$ dS/m is reached throughout the 50-year horizon for the control site irrigated with well water and after 8, 9 and 14 years of irrigated with blended wastewater for sites 2, 6, and 7 respectively. For sites 3, 4, and 5 soil water EC increased above 2.0 dS/m in the simulation period. Actual ET had a positive and significant association whereas irrigation amounts had a negative and significant association with $EC_{sw} \leq 2.0$ dS/m.

While we believe that the modeling results can inform recommendations about irrigation management practices and for estimating salt output loading resulting from use of saline waters for irrigation, difficulties in linking field observations of soil salinity and model predictions remain troubling. However, since we account for irrigation water salinity, rainfall salinity, and dissolution of salts in the soil and exclude additions of fertilizer and soil amendment our simulated EC_{sw} values are likely a lower bound of soil salinity associated with the irrigation and farm management practices considered in the modeling. Nonetheless, it is important to note that this model is user-friendly and less data intensive and it can be very useful for setting reference benchmarks of long-term salinity impacts of using saline water for irrigation.

Author Contributions: Conceptualization, P.Z. and M.G.; Data curation, P.Z. and M.G.; Formal analysis, P.Z.; Funding acquisition, M.G.; Investigation, M.G.; Methodology, P.Z., I.K. and M.G.; Project administration, M.G.; Resources, M.G.; Software, P.Z.; Supervision, I.K. and M.G.; Validation, P.Z. and M.G.; Visualization, P.Z.; Writing—Original draft, P.Z. and M.G.; Writing—Review & editing, I.K. and M.G.

Funding: The Monterey County Water Resources Agency funded the water and soil sampling monitoring program and collection of cropping data.

Acknowledgments: We thank Stephen Grattan and Daniel Isidoro for providing notes and equations for the root zone salinity model and Belinda Platts and the Monterey County Water Resources Agency for water and soil sampling and crop data.

Conflicts of Interest: The authors declare no conflict of interest.

Appendix A

Table A1. Cropping patterns for the control site and sites irrigated with treated wastewater.

Site #	Cropping Pattern	Crop	Planting Month	Harvest Month	Average Growing Days
Control	Lettuce, Broccoli, Cauliflower, Cabbage	Lettuce	Mar–Aug	Jun–Nov	72
		Broccoli	Jul–Aug	Oct–Dec	101
		Cauliflower	Jul–Nov	Apr–Oct	118
		Cabbage	Apr	Jul	98
2	Lettuce, Broccoli, Cauliflower, Spinach, Celery	Lettuce	Jan–Sep	Apr–Nov	74
		Broccoli	Jan–Jun	May–Oct	104
		Cauliflower	May–Aug	Aug–Nov	94
		Spinach	Sep	Oct	49
		Celery	Jul	Oct	93
3	Lettuce, Broccoli, Cauliflower, Strawberry	Lettuce	Mar–Jul	May–Oct	73
		Broccoli	Feb–Jul	Jun–Oct	104
		Cauliflower	Feb–Apr	May–Jul	94
		Strawberry	Nov	Oct–Nov	344
4 & 5	Artichoke	1st crop		May	Annual
		2nd crop		Oct–Nov	
6	Lettuce, Broccoli, Cauliflower, Strawberry, Celery	Lettuce	Jan–Jul	Apr–Sep	74
		Broccoli	Apr–Jul	Jul–Oct	92
		Cauliflower	Jan–Jul	May–Nov	99
		Strawberry	Nov	Nov	344
		Celery	May–Jul	Aug–Oct	95
7	Lettuce, Cauliflower, Broccoli	Lettuce	Mar–Aug	May–Oct	72
		Cauliflower	Apr–Aug	Jul–Aug	92
		Broccoli	Jul	Oct	97

Table A2. Time-averaged crop coefficients and maximum rooting depth.

Crop	K_c Values			Rooting Depth (cm)
	Initial	Midseason	Late	
Artichoke	0.5	1	0.95	90
Broccoli	0.7	1.05	0.95	60
Cauliflower	0.7	1.05	0.95	70
Celery	0.7	1.05	1	50
Lettuce	0.7	1	0.95	50
Spinach	0.7	0.9	0.95	50
Strawberry	0.4	0.9	0.85	30

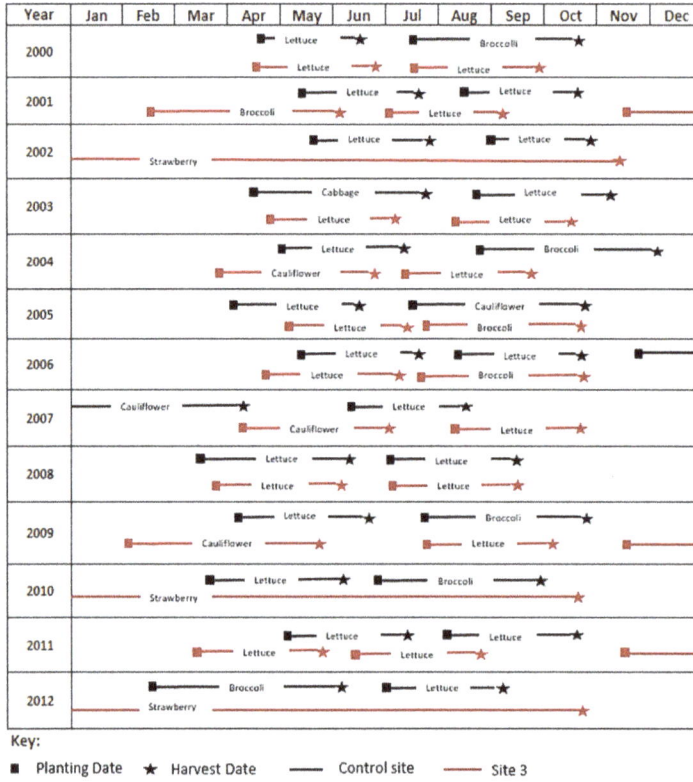

Figure A1. Crop rotation schedule for the control site and site 3.

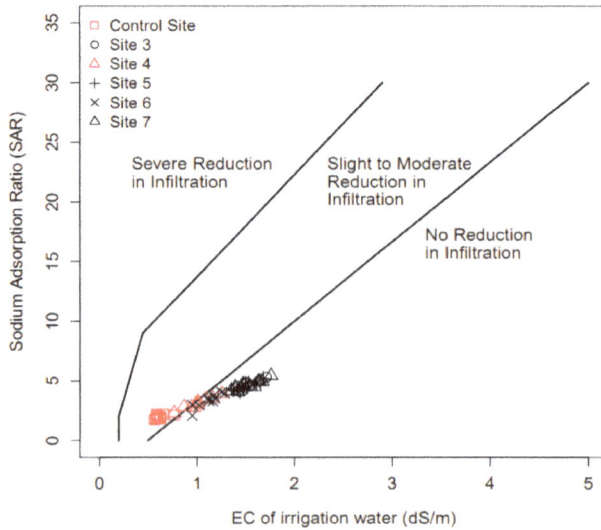

Figure A2. Effect of salinity and sodium adsorption ratio of irrigation water on infiltration rate. (Modified from Reference [6] (p. 44)).

Appendix B

Unsaturated Soil Water Movement

The soil matric potential (ψ) was related to the volumetric water content (θ) by means of Equation (A1) [32]:

$$\varphi = \varphi_s \times \left(\frac{\theta}{\theta_s}\right)^{-b} \tag{A1}$$

Where θ_s was the volumetric water content at saturation, ψ_s is the water entry potential or "saturation" water potential and b is the slope of the water retention curve on a logarithmic plot. For each soil type, b and ψ_s were calculated from the volumetric water content at field capacity and wilting point and their respective potentials in absolute value (ψ_{FC} = 316 cm and ψ_{WP} = 15,849 cm; so that pF (FC) = 2.5 and pF (WP) = 4.2). Taking logarithms, the expression of the potentials for FC and WP become linear equations:

$$\log(\varphi_{FC}) = 2.5 = \log\varphi_s - b.\log\left(\frac{\theta_{FC}}{\theta_s}\right)$$
$$\log(\varphi_{WP}) = 4.2 = \log\varphi_s - b.\log\left(\frac{\theta_{WP}}{\theta_s}\right) \tag{A2}$$

from which, b and ψ_s are estimated. The unsaturated hydraulic conductivity for a given θ was given by:

$$K = K_s \times \left(\frac{\theta}{\theta_s}\right)^{2b+3} \tag{A3}$$

where K_s is the saturated hydraulic conductivity. Thus, unsaturated flow between layers (U) can be calculated as:

$$U = K\left(\frac{\Delta\varphi}{\Delta Z}\right) \tag{A4}$$

where ΔZ is the center to center simulation distance selected between layers and neglecting the gravitational gradient.

Crop Water Uptake

Non-stressed crop ET is calculated as:

$$ET_c = K_c \times ET_o \tag{A5}$$

where K_c is the crop coefficient and varies with the crop development stages (Table A2) and ET_o is the reference ET. Between cropping seasons, all ET or evaporation E was assumed to take place from the upper layer. For this period Kc was calculated from the mean interval between precipitation events of each month and the mean precipitation event in each month and ET_o [38].

In each layer (k), the actual crop ET can be lower than ET_c(k) due to water stress, which depends on the soil water content and the sensitivity of the crop to low water contents, accounted for through the crop-specific parameter p: the ratio of readily available soil water (RAW) to total available water (TAW) (p = RAW/TAW) [38]. When the soil water content (W(k)) in a layer fell below We(k) = WP + (1 − p) TAW, the ET from that layer actual crop ET(k) dropped below the ETc(k), and the actual ET of the layer was calculated as:

$$\text{actual crop ET} = K_s \times ET_c \tag{A6}$$

where K_s is a stress coefficient [38]:

$$K_s = \begin{array}{ll} 1 & \text{if } W(k) > W_e(k) \\ \frac{W(k) - W_e(k)}{W_e(k) - WP(k)} & \text{if } WP(k) < W(k) < W_e(k) \\ 0, & \text{if } W(k) < WP(k) \end{array} \tag{A7}$$

when one layer was stressed during the growing season ($W(k) < W_e(k)$), the model allowed increase in the extraction coefficient of the lower layer to supply the ET demand of the day. The root zone water uptake pattern depends on irrigation frequency. Root uptake patterns were taken from [18,42,43]. Root length increase a function time is calculated as [45]:

$$L_z = L_o + (L_{max} - L_0) \times \sqrt{\frac{(t - \frac{t_0}{2})}{(t_{L_{max}} - \frac{t_0}{2})}} \tag{A8}$$

where L_z is the rooting depth at time t, L_o is the starting root depth, L_{max} is the maximum root length, $t_{L_{max}}$ time after planting when L_{max} is reached and t_o is time to reach 90% crop emergence. This is a linear root expansion; the method assumes that once half of the time required for crop emergence is passed by $\frac{t_0}{2}$, the rooting depth starts to increase from an initial depth L_o till L_{max} is reached.

Water Balance

Surface runoff for winter rainfall and a fraction of applied water is modelled using the SCS method. We define curve number (CN) associated with row crop cover for the growing season and bare soil for non-growing season from the SCS tables and calculate precipitation runoff as:

$$SR(p) = \begin{array}{ll} 0 & \text{if } P \leq 0.2S \\ \frac{(P - 0.2S)^2}{P + 0.8S} & \text{if } P > 0.2S \end{array} \tag{A9}$$

where P is runoff producing precipitation and S is the potential maximum retention after runoff begins related to CN by:

$$S = 254 \times (\frac{100}{CN} - 1) \tag{A10}$$

Daily water balance for the 4-layered root zone and 2-layered vadose zone is performed. To account for the slow water movement between layers for low water content below field capacity, a slow upward or downward flow U is calculated dependent upon the difference in matric potential between soil layers (Equations B4). In the first quarter of the root zone inflows and outflows include applied water (I) and rainfall (P), the drainage above field capacity (D (1)) to layer 2, actual crop ET and U. For the underlying root zone layers, inflows and outflows include drainage (D) from the overlying layer, U and actual crop ET and finally for the unsaturated layers below the root zone inflows and outflows include D and U.

When the soil water content in layer "k" is above field capacity, the excess water drains to the lower layer over a two-day period, the higher flow in the first day than the second. The fraction α of the excess water that drains the first day is calculated from the soil texture in the layer through an empirical relation obtained to match results presented by approximately 0.9 for sand, 0.85 for loam and 0.7 for clay [49]. Two arbitrary water contents were defined from field capacity to saturation for each layer W_a and W_b defined as:

$$\begin{array}{l} W_a = (1 - \alpha) \times (W_s - W_{FC}) + W_{FC} \\ W_b = (1 - \alpha) \times (W_S - W_{FC}) + W_{FC} \end{array} \tag{A11}$$

Drainage (D) is calculated as:

$$D = \begin{array}{ll} \alpha \times (W_s - W_{FC}) & \text{if } W > W_b \\ (1 - \alpha) \times (W_s - W_{FC}) & \text{if } W_a < W < W_b \\ W - W_{FC} & \text{if } W < W_a \end{array} \tag{A12}$$

where W_{FC} and W_s define field capacity and saturation of a soil layer. After taking out actual crop ET and D outflows from the layer, we also need to account for upwards or downward movement of water (U) dependent upon the difference in matric potential between soil layers (Equation (A4)).

Salt Balance

Salt balance was performed in conjunction with the water balance assuming complete mixing of water entering each layer with that already stored in that layer. The electrical conductivity of water (EC) was used as an indicator of salinity, assuming implicitly that there was a unique relationship between EC and total dissolved solids (TDS) in these dilute solutions and that the EC behaves like a non-reactive solute. Salinity of the input waters (irrigation water (EC_w) and precipitation (EC_p)) must be known. The mass of salts in layer k ($Z(k)$) is estimated from the product $EC_{sw}(k) W(k)$, where EC_{sw} is the electrical conductivity of the soil water in that layer. The mass of salts in layer k in day 1 (t + 1) results from the salinity in day 0 (or t) and the salt fluxes in day 1 that are added sequentially. Accounting for layer 1 for example is as follows: salts in I, P, and mineral dissolution (k_d) are added to the salt mass in layer 1 to obtain $Z^a(1)_1$:

$$Z^a(1)_1 = Z(1)_0 + EC_w \times I_1 + EC_p \times P_1 + k_d \tag{A13}$$

This results in a soil water concentration of:

$$EC_{sw}^a(1)_1 = Z^a(1)_1 / (W(1)_0 + I_1 + P_1) \tag{A14}$$

Drainage takes place with concentration EC_{sw}^a so that the new mass of salts is:

$$Z^b(1)_1 = Z^a(1)_1 - EC_{sw}^a(1)_1 \times D(1)_1 \tag{A15}$$

and the new soil water concentration is:

$$EC_{sw}^b(1)_1 = Z^b(1)_1 / (W(1)_t + I_1 + P_1 - D(1)_1) \tag{A16}$$

The soil at this state is evapo-concentrated by crop water uptake (actual crop ET):

$$EC_{sw}^c(1)_1 = Z^b(1)_1 / (W(1)_t + I_1 + P_1 - D(1)_1 - \text{actual crop ET}(1)_1) \tag{A17}$$

The mass of salts in the slow flow U are then added or removed to obtain the final mass of salts in the layer:

$$Z(1)_1 = \begin{array}{ll} Z^b(1)_1 - U_{1-2} \times EC_{sw}^c(2)_1 & \text{if } U_{1-2} < 0 \\ Z^b(1)_1 - U_{1-2} \times EC_{sw}^c(1)_1 & \text{if } U_{1-2} > 0 \end{array} \tag{A18}$$

which allows for calculating the final soil water concentration:

$$EC_{sw}(1)_1 = Z(1)_1 / W(1)_1 \tag{A19}$$

References

1. Hillel, D. *Salinity Managment for Sustainable Irrigation: Integrating Science, Environment, and Economics*; The World Bank: Washington, DC, USA, 2000.
2. USBR. *Quality of Water—Colorado River Progress Report No. 24*; U.S. Bureau of Reclamation: Salt Lake City, UT, USA, 2013.
3. Rose, D.A.; Konukcu, F.; Gowing, J.W. Effect of water table depth on evaporation and salt accumulation from saline groundwater. *Aust. J. Soil Res.* **2005**, *43*, 565–573. [CrossRef]
4. Grismer, M.E.; Gates, T.K. Estimating saline water table contributions to crop water use. *Calif. Agric.* **1988**, *42*, 23–24.

5. Ragab, R.A.; Amer, F. Estimating water table contribution to the water supply of maize. *Agric. Water Manag.* **1986**, *11*, 221–230. [CrossRef]
6. Hanson, B.R.; Grattan, R.R.; Fulton, A. *Agricultural Salinity and Drainage*; University of California, Davis: Davis, CA, USA, 2006.
7. Konikow, L.F.; Rielly, T.E. Seawater intrusion in the United States. In *Seawater Intrution in Coastal Aquifers*; Bear, J., Cheng, A.H.D., Sorek, S., Ouazar, D., Herrera, I., Eds.; Springer: dordrecht, The Netherlands, 1999; pp. 463–506.
8. Platts, B.E.; Grismer, M.E. Chloride levels increase after 13 years of recycled water use in the Salinas Valley. *Calif. Agric.* **2014**, *68*, 7. [CrossRef]
9. MCWRA. *State of the Salinas River Groundwater Basin*; Monterey County Water Resources Agency: Salinas, CA, USA, 2015.
10. Vengosh, A.; Gill, J.; Davisson, L.M.; Hudson, G.B. A multi-isotope (B, Sr, O, H, and C) and age dating (^{3}H–^{3}He and ^{14}C) study of groundwater from Salinas Valley, California: Hydrochemistry, dynamics, and contamination processes. *Water Resour. Res.* **2002**, *38*, 9-1–9-17. [CrossRef]
11. Platts, B.; Grismer, M.E. Rainfall leaching is critical for long-term use of recycled water in the Salinas Valley. *Calif. Agric.* **2014**, *68*, 75–78. [CrossRef]
12. Letey, J.; Hoffman, G.J.; Hopmans, J.W.; Grattan, S.R.; Suarez, D.; Corwin, D.L.; Oster, J.D.; Wu, L.; Amrhein, C. Evaluation of soil salinity leaching requirement guidelines. *Agric. Water Manag.* **2011**, *98*, 502–506. [CrossRef]
13. SHI National Soil Health Measurements to Accelerate Agricultural Transformation. Available online: http://soilhealthinstitute.org/national-soil-health-measurements-accelerate-agricultural-transformation/ (accessed on 23 August 2017).
14. Maas, E.V.; Hoffman, G.J. Crop Salt Tolerance. *J. Irrig. Drain.* **1977**, *103*, 20.
15. Rhoades, J.D.; Kandiah, A.; Mashali, A.M. *The Use of Saline Waters for Crop Production*; FAO: Rome, Italy, 1992; Volume FAO Irrigation & Drainage, p. 48.
16. Shelia, V.; Simunek, J.; Boote, K.; Hoogenboom, G. Coupling DSSAT and HYDRUS-1D for simulations of soil water dynamics in the soil-plant-atmosphere system. *J. Hydrol. Hydromech.* **2018**, *66*, 232–245. [CrossRef]
17. Isidoro, D.; Grattan, S.R. Predicting soil salinity in response to different irrigation practices, soil types and rainfall scenarios. *Irrig. Sci.* **2011**, *29*, 197–211. [CrossRef]
18. Ayers, R.S.; Westcot, D.W. *Water Quality for Agriculture*; Food and Agriculture Organization of the United Nations: Rome, Italy, 1985.
19. Gates, T.K.; Grismer, M.E. Stochastic approximation applied to optimal irrigation and drainage planning. *J. Irrig. Drain.* **1989**, *115*, 255–283. [CrossRef]
20. Grismer, M.E. Pore-size distribution and infiltration. *Soil Sci.* **1986**, *141*, 249–260. [CrossRef]
21. Ayars, J.; Christen, E.; Soppe, R.; Meyer, W. The resource potential of in-situ shallow groundwater use in irrigated agriculture. *Irrig. Sci.* **2006**, *24*, 147–160. [CrossRef]
22. Grismer, M.E. *Use of Shallow Groundater for Crop Production*; UC Agriculture & Natural Resources: Davis, CA, USA, 2015; pp. 1–6.
23. Grismer, M.E.; Bali, K.M. Subsurface drainage systems have little impact on water tables salinity of clay soils. *Calif. Agric.* **1998**, *52*, 18–22. [CrossRef]
24. Grismer, M.E.; Gates, T.K.; Hanson, B.R. Irrigation and drainage strategies in saline problem areas. *Calif. Agric.* **1988**, *42*, 23–24.
25. Talsma, T. *The Control of Saline Groundwater*; Meded, Landbouwhogeschool: Wageningen, The Netherlands, 1963; Volume 63, pp. 1–68.
26. Durbin, T.J.; Kapple, G.W.; Freckleton, J.R. *Two-Dimensional and Three-Dimensional Digital Flow Models of the Salinas Valley grouNd-Water Basin, California*; Water Resources Division, US Geological Survey: Reston, VA, USA, 1978.
27. Hall, P. *Selected Geological Cross Sections in the Salinas Valley Using GeoBASE*; Monterey County Water Resources Agency: Salinas, CA, USA, 1992.
28. Fogg, G.E.; Labolle, E.M.; Weissmann, G.S. Groundwater Vulnerability Assessment: Hydrogeologic Perspective and Example from Salinas Valley, California. In *Assessment of Non-Point Source Pollution in the Vadose Zone*; American Geophysical Union: Washington, DC, USA, 2013; pp. 45–61.
29. University of California, Division of Agriculture and Natural Resources. SoilWeb. Available online: https://casoilresource.lawr.ucdavis.edu/gmap/ (accessed on 7 October 2018).

30. Šimůnek, J.; Šejna, M.; Saito, H.; Sakai, M.; van Genuchten, M.T. *The Hydrus-1D Software Package for Simulating the Movement of Water, Heat, and Multiple Solutes in Variably Saturated Media*; 4.14; Department of Environmental Sciences, University of California Riverside: Riverside, CA, USA, 2013.

31. DWR CIMIS: California Irrigation Managemetn Information System. Available online: https://cimis.water. ca.gov/Resources.aspx (accessed on 6 April 2017).

32. Clapp, R.B.; Hornberger, G.M. Empirical equations for some soil hydraulic properties. *Water Resour. Res.* **1978**, *14*, 601–604. [CrossRef]

33. Brooks, R.H.; Corey, A.T. Hydraulic properties of porous media. In *Hydrology Papers*; Colorado State University: Fort Collins, CO, USA, 1964.

34. Gardner, W.R. Some steady state solutions of unsaturated moisture flow equations with application to evaporation from a water table. *Soil Sci.* **1958**, *84*, 228–232. [CrossRef]

35. Haverkamp, R.; Vaclin, M.; Touma, J.; Wierenga, P.J.; Vachaud, G. A comparison of numerical simulation models for one-dimensional infiltration. *Soil Sci. Soc. Am. J.* **1977**, *41*, 285–294. [CrossRef]

36. Mualem, Y. A new model for predicting the hydraulic conductivity of unsaturated porous media. *Water Resour. Res.* **1976**, *12*, 513–522. [CrossRef]

37. van Genuchten, M.T. A closed-form equation for predicting the hydraulic conductivity of unsaturated soils. *Soil Sci. Soc. Am. J.* **1980**, *44*, 892–898. [CrossRef]

38. Allen, R.G.; Pereira, L.S.; Raes, D.; Smith, M. Crop evapotranspiration-Guidelines for computing crop water requirements-FAO Irrigation and drainage paper 56. *FAO Rome* **1998**, *300*, D05109.

39. Hjelmfelt, A.T. Investigation of curve number procedure. *J. Hydraul. Eng.* **1991**, *117*, 725–737. [CrossRef]

40. USDA. *Natural Resources Conservation Service National Engineering Handbook*; USDA: Washington, DC, USA, 2008; Volume Part 623 Section 15.

41. Raats, P.A.C. Distribution of salts in the root zone. *J. Hydrol.* **1975**, *27*, 237–248. [CrossRef]

42. Rhoades, J.D. Use of saline drainage water for irrigation. In *Agricultural Drainage*; Skaggs, R.W., van Schilfgaarde, J., Eds.; Agronomy Monograph, ASA-CSSA-SSSA: Madison, WI, USA, 1999; Volume 38, pp. 615–657.

43. Skaggs, T.H.; Anderson, R.G.; Corwin, D.L.; Suarez, D.L. Analytical steady-state solutions for water-limited cropping systems using saline irrigation water. *Water Resour. Res.* **2014**, *50*, 9656–9674. [CrossRef]

44. Grattan, S.R.; Grieve, C.M. Salinity–mineral nutrient relations in horticultural crops. *Sci. Horticult.* **1998**, *78*, 127–157. [CrossRef]

45. Steduto, P.; Hsiao, T.C.; Raes, D.; Fereres, E. *Crop yield Response to Water*; Food and Agriculture Organization of the United Nations Rome: Rome, Italy, 2012.

46. Orang, M.N.; Snyder, R. *Consumptive Use Program-CUP+*; California Department of Water Resources: Sacramento, CA, USA, 2013.

47. Cahn, M. Estimated Crop Coefficients for Strawberry. In *Salinas Valley Agriculture*; UCANR: Salinas Valley, CA, USA, 2012; Volume 2017.

48. Herman, J.; Usher, W. SALib: An open-source Python library for sensitivity analysis. *J. Open Source Softw.* **2017**, *2*, 97. [CrossRef]

49. Hillel, D.; van Bavel, C.H.M. Simulation of Profile Water Storage as Related to Soil Hydraulic Properties. *Soil Sci. Soc. Am. J.* **1976**, *40*, 807–815. [CrossRef]

50. DWR. *California's Groundwater Bulletin 118*; California Department of Water Resources: Sacramento, CA, USA, 2004.

51. Saltelli, A.; Ratto, M.; Andres, T.; Campolongo, F.; Cariboni, J.; Gatelli, D.; Saisana, M.; Tarantola, S. *Global Sensitivity Analysis: The Primer*; Wiley: Hoboken, NJ, USA, 2008.

52. Morris, M.D. Factorial sampling plans for preliminary computational experiments. *Technometrics* **1991**, *33*, 161–174. [CrossRef]

53. Campolongo, F.; Cariboni, J.; Saltelli, A. An effective screening design for sensitivity analysis of large models. *Environ. Model. Softw.* **2007**, *22*, 1509–1518. [CrossRef]

54. Sobol', I.M. Global sensitivity indices for nonlinear mathematical models and their Monte Carlo estimates. *Math. Comput. Simul.* **2001**, *55*, 271–280. [CrossRef]

55. Saltelli, A. Making best use of model evaluations to compute sensitivity indices. *Comput. Phys. Commun.* **2002**, *145*, 280–297. [CrossRef]

56. McKay, M.; Beckman, R.; Conover, W. A comparison of three methods for selecting values of input variables in the analyss of output from a computer code. *Technometrics* **1979**, *21*, 239–245.
57. O'Geen, A. Reclaiming Saline, Sodic, and Saline-Sodic Soils. In *Drought Tip*; University of California, Agriculture and Natural Resources: Richmond, CA, USA, 2015.
58. Mullen, K.; Ardia, D.; Gil, D.; Windover, D.; Cline, J. DEoptim: An R Package for Global Optimization by Differential Evolution. *J. Stat. Softw.* **2011**, *40*, 1–26. [CrossRef]
59. Hanson, B.; Hopmans, J.W.; Simunek, J. Leaching with Subsurface Drip Irrigation under Saline, Shallow Groundwater Conditions. *Vadose Zone J.* **2008**, *7*, 810–818. [CrossRef]
60. McLeod, A.I. *Kendall*, version 2.2; Univerity of Western Ontario: London, ON, Canada, 2015.
61. Grieve, C.M.; Grattan, S.R.; Maas, E.V. Plant salt tolerence. In *Angricultural Salinity Assessment and Management*, 2nd ed.; Wallender, W.W., Tanji, K.K., Eds.; American Society of Civil Engineers: Reston, VA, USA, 2012; pp. 405–459.

agriculture

MDPI

Article

Research Advances in Adopting Drip Irrigation for California Organic Spinach: Preliminary Findings

Ali Montazar [1],*, Michael Cahn [2] and Alexander Putman [3]

[1] Division of Agriculture and Natural Resources, University of California, UCCE Imperial County,
    1050 East Holton Road, Holtville, CA 92250, USA
[2] Division of Agriculture and Natural Resources, University of California, UCCE Monterey County,
    1432 Abbott Street, Salinas, CA 93901, USA
[3] Department of Microbiology and Plant Pathology, University of California, Riverside,
    900 University Avenue, Riverside, CA 92521, USA
* Correspondence: amontazar@ucanr.edu

Received: 13 June 2019; Accepted: 6 August 2019; Published: 9 August 2019

Abstract: The main objective of this study was to explore the viability of drip irrigation for organic spinach production and the management of spinach downy mildew disease in California. The experiment was conducted over two crop seasons at the University of California Desert Research and Extension Center located in the low desert of California. Various combinations of dripline spacings and installation depths were assessed and compared with sprinkler irrigation as control treatment. Comprehensive data collection was carried out to fully understand the differences between the irrigation treatments. Statistical analysis indicated very strong evidence for an overall effect of the irrigation system on spinach fresh yields, while the number of driplines in bed had a significant impact on the shoot biomass yield. The developed canopy crop curves revealed that the leaf density of drip irrigation treatments was slightly behind (1–4 days, depending on the irrigation treatment and crop season) that of the sprinkler irrigation treatment in time. The results also demonstrated an overall effect of irrigation treatment on downy mildew, in which downy mildew incidence was lower in plots irrigated by drips following emergence when compared to the sprinkler. The study concluded that drip irrigation has the potential to be used to produce organic spinach, conserve water, enhance the efficiency of water use, and manage downy mildew, but further work is required to optimize system design, irrigation, and nitrogen management practices, as well as strategies to maintain productivity and economic viability of utilizing drip irrigation for spinach.

Keywords: downy mildew; drip irrigation; irrigation management; organic production; spinach

1. Introduction

Spinach (*Spinacia oleracea* L.) is a fast-maturing, cool-season vegetable crop. In California, spinach is mainly produced in four areas of the southern desert valleys, the southern coast, the central coast, and the central San Joaquin Valley. The farm gate value of Californian non-processing spinach from a total planted area of 13,557 ha was nearly 242 million dollars in 2017 [1]. The area in California under spinach production increased more than 30% over the last 10 years; the planted area increased from 1478 ha in 2008 to 3893 ha in 2017 in the Imperial Valley [2,3].

Water and nitrogen are generally essential drivers for plant growth and survival, and for spinach, are two key factors that considerably affect yield and quality [4]. Spinach is highly responsive to nitrogen fertilization [5] and accumulates as much as 134 kg ha^{-1} of nitrogen in 30 days [6]. Spinach yield was shown to increase as a result of high nitrogen application rates but decreased when nitrogen application rate was excessive [6]. Researchers reported that with the increase of nitrogen fertilizer application, nitrogen use efficiency of spinach was significantly decreased; however, water use efficiency (as the ratio of yield to the seasonal crop water use) of spinach was increased in most cases [5,7]. Reducing nitrogen fertilizer use in spinach will be a challenge. The crop has a shallow root system, a high N demand, which occurs over a short period, and strict quality standards for a deep green color that tends to encourage N applications beyond the agronomic requirements to maximize yield [6]. Addressing these challenges requires that nitrogen fertilizer is applied at the optimal time and rate based on crop uptake and the soil–nitrogen test, and that irrigation water is efficiently applied to minimize leaching. The unpublished data from a survey conducted by the University of California Cooperative Extension Monterey county indicates that spinach-growers generally apply more water than that needed by crop water (100% to 150% more).

Downy mildew on spinach is a widespread and very destructive disease in California. It is a disease that is the most significant issue facing the spinach industry, and crop losses can be significant in all areas where spinach is produced [8]. *Peronospora effusa* is an obligate oomycete that causes downy mildew of spinach. Spinach growers rely on host resistance and fungicides to manage downy mildew. This dependence on resistance was relatively effective until recently, when malignant forms of *P. effusa* began to emerge in rapid succession [9]. In recent years, several new downy mildew races have appeared in the state of California, raising concerns about the ability to manage this threat and causing the industry to consider research strategies to address the problem. Organic spinach producers are especially vulnerable to these virulent strains because synthetic fungicide use is prohibited, and choice in regard to variety is determined at planting. This has led to significant yield losses in organic spinach production.

Like all downy mildew pathogens, *P. effusa* requires cool and wet conditions for infection and disease development [8,10]. The California industry is known for using very high planting densities and a large number of seed lines per bed [10]. For the baby leaf or clipped markets, the planting density is usually 8.6–9.8 million seeds per hectare [11]. The dense canopy of spinach retains much moisture, and creates ideal conditions for infection and disease development. Spores (called sporangia) are dispersed at short distances via wind or splashing water, or at medium distances via wind. In addition, most conventional and organic spinach fields are irrigated by sprinkler irrigation in California. Overhead irrigation deposits free moisture on leaf surfaces and increases relative humidity in the canopy, which contributes to the speed and severity of downy mildew epidemics within a field when other conditions, such as temperature, are favorable. Production practices that reduce the favorability of the spinach canopy for downy mildew development are needed to reduce losses to downy mildew in both organic and conventional production.

Because most conventional and organic spinach fields are irrigated by sprinkler irrigation in California, overhead irrigation could contribute to the speed and severity of downy mildew epidemics within a field when other conditions, such as temperature, are favorable. While preventing downy mildew on spinach is of high interest of organic vegetable production in California, integrated approaches are greatly required to reduce yield losses from this major disease. New irrigation management techniques and practices in spinach production may have a significant economic impact to the leafy greens industry through the control of downy mildew. In addition to losses from plant pathogens, new irrigation practices could reduce risks to food safety caused by overhead application of irrigation water.

Drip irrigation has revolutionized crop production systems in western states of the US by increasing yields and water-use efficiency in many crops [12]. Sub-surface drip irrigation was successfully implemented on vegetable crops, such as processing tomato, where yields have increased by 20–50% over recent years since this practice was adapted [13]. While sub-surface drip irrigation has been utilized primarily for high-value specialty crops, several studies have reported the benefits of its application for agronomic low-value crops [12–16]. Drip irrigation was evaluated versus micro-sprinkler irrigation in spinach [17]. The drip was found to be better than the micro-sprinkler because of greater yields and the lower installation cost. The effectiveness of drip fertigation for reducing nitrate in spinach was reported by Takebe et al. [18]. Drip fertigation was considered to reduce nitrate more stably. In addition to higher yield, sub-surface and surface drip irrigation can reduce risks of plant diseases in vegetables by minimizing leaf wetness and waterlogged soil conditions.

Adapting drip irrigation for high-density spinach plantings may be a possible solution to reduce losses from downy mildew, improve crop productivity and quality, and conserve water and fertilizer. Currently, drip irrigation is not used for producing spinach in California, and there is a lack of information on the viability of this technology and optimal practices for irrigating spinach with drips. In fact, to our knowledge, few studies have been conducted to assess the potential benefit of drip irrigation for this high-density planted spinach. As an initial test, the main objective of this project was to evaluate the viability of adapting drip irrigation for organic spinach production. The project was particularly aimed at understanding the system design to successfully produce spinach, and to conduct a preliminary assessment on the impact of drip irrigation on the management of spinach downy mildew.

2. Materials and Methods

The field experiments were carried out over two crop seasons at the organic field of the University of California Desert Research and Extension Center (UC DREC) (32°48′35″ N; 115° 26′39″ W; 22 m below mean sea level) located in the Imperial Valley, California. The field had a silty clay soil (the top 30 cm soil surface contains 14% sand, 42% silt, and 44% clay). Soil characteristics referring to three genetic horizons selected from the soil survey are presented in Table 1. Soil pH ranged between 7.94 and 8.04, and its electric conductivity was from 1.3 to 2.9 dS m^{-1}. The area has a desert climate with a mean annual rainfall of 76 mm and air temperature of 22 °C. Spinach can typically grow from October to March in this region. The average pH and electrical conductivity of water supply (surface water from the Colorado River) was 8.0 and 1.18 ds/m during the experiment, respectively.

Table 1. Soil characteristics of the study site.

Soil Parameters	Soil Depth		
Generic horizon (cm)	0–30	30–60	60–90
Texture	silty clay	clay loam	clay
Sand (%)	14	21	8
Silt (%)	42	41	34
Clay (%)	44	38	58
pH	7.94	8.04	8.02
EC (dS m^{-1})	1.3	1.4	2.9
Cation exchange capacity (meq/1000 g)	48.6	50.7	54.0
Bulk density (g cm^{-3})	1.54	1.48	1.45
Exchange sodium percentage (%)	3.5	4.1	6

2.1. Experimental Design and Treatments

Experiment 1 (fall experiment): Land preparation was conducted in late September 2018, and untreated Viroflay spinach seeds were planted at a rate of 37 kg ha^{-1} on 31st October. For the research trial (Figure 1a), five irrigation system treatments consisted of two drip depths (driplines on the soil surface and driplines at 3.8 cm depth), two dripline spacings (three driplines on a 203 cm bed and four driplines on an 203 cm bed), and sprinkler irrigation (203 cm bed). The experiment was arranged in a randomized complete block with four replications. Each drip replication had three beds, and each sprinkler replication had six beds. The beds were 61 m long. Figure 2 presents the individual experimental beds with four driplines at 3.8 cm and on the soil.

All treatments were germinated by sprinklers (two sets of five-hour irrigations). A flow control drip tape from the Toro company was used with a hose diameter of 15.88 mm, wall thickness of 0.154 mm, emitter spacing of 20.3 cm, and operating emitter flowrate of 0.49 L/h using the Nelson sprinklers R200WF Rotator (280 L/h @ 3.45 Bar) were used with spacings of 12.2 m × 9.1 m. Water distribution uniformity was measured for the sprinkler irrigation using the ASABE (American Society of Agricultural and Biological Engineers) standard method [19]. The average distribution uniformity for the sprinkler irrigation was 80%, and an average of 92% was assumed as water distribution uniformity of the drip irrigation.

True 6-6-2 (a homogeneous pelleted fertilizer from True Organic Products) was applied at a rate of 89 kg of N per hectare as pre-plant fertilizer, and True 4-1-3 (a liquid fertilizer from True Organic Products) was applied as complementary fertilizer through injection into the irrigation system. For the drip treatments, True 4-1-3 was applied three times after germination (by crop harvest) at a rate of 45, 33, and 44 kg of N per hectare. This liquid fertilizer was applied at a rate of 56, 42, and 50 kg of N per hectare for the sprinkler irrigation system. Soil tests conducted before planting were used to determine the status of available nitrogen/nutrients to develop fertilizer recommendations to achieve optimum crop production and prevent nitrogen deficiency. Initially, the crop was irrigated with more water than required, as determined by following crop evapotranspiration (ET, as the sum of soil evaporation and plant transpiration), and using soil moisture data, we tried to irrigate spinach trials more than crop water requirements to make sure there was no water stress for the entire crop season. However, according to our data, this led to over-irrigation at some points in the early and mid-crop season. Irrigation events were typically scheduled once a week for the sprinkler treatments and twice a week for the drip treatments at the required running times of each system/treatment.

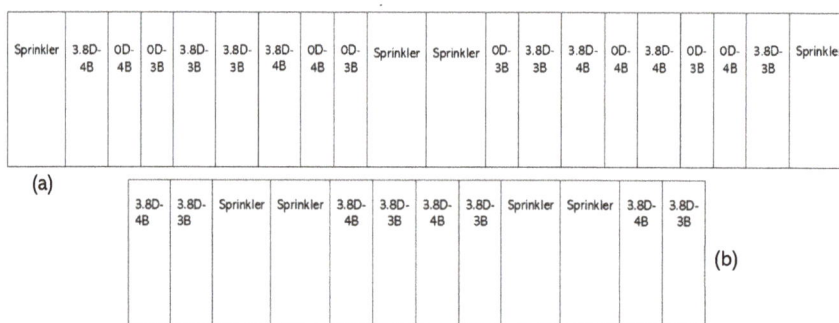

Sprinkler	3.8D-4B	OD-4B	OD-3B	3.8D-3B	3.8D-3B	3.8D-4B	OD-4B	OD-3B	Sprinkler	Sprinkler	OD-3B	3.8D-3B	3.8D-4B	OD-4B	3.8D-4B	OD-3B	OD-4B	3.8D-3B	Sprinkler

(a)

3.8D-4B	3.8D-3B	Sprinkler	Sprinkler	3.8D-4B	3.8D-3B	3.8D-4B	3.8D-3B	Sprinkler	Sprinkler	3.8D-4B	3.8D-3B

(b)

Figure 1. The fall (**a**) and winter (**b**) experiment layouts (not to scale). Sprinkler: treatment irrigated by sprinklers; 3.8D-3B: treatment with three driplines in each bed installed at 3.8 cm depth; 3.8D-4B: treatment with four driplines in each bed installed at 3.8 cm depth; 0D-4B: treatment with four driplines in each bed on the soil surface; 0D-3B: treatment with three driplines in each bed on the soil surface.

(a)　　　　　　　(b)

Figure 2. Beds with four driplines in the fall crop season. The pictures demonstrate individual beds with four driplines at 3.8 cm depth 30 days after planting (**a**) and on the soil surface 12 days after planting (**b**).

Experiment 2 (winter experiment): Untreated Viroflay spinach seeds were planted at a rate of 38 kg ha^{-1} on 28th January. For the research trial (Figure 1b), three irrigation system treatments consisted of two dripline spacings (three driplines on a 203 cm bed and four driplines on a 203 cm bed), and sprinkler irrigation (203 cm bed). All driplines were installed at 3.8 cm depth. The experiment was arranged in a randomized complete block with four replications. Each drip and sprinkler treatment replication consisted of three beds × 61 m length. All treatments were germinated by sprinklers (two sets of five hours). The buffer beds located between sprinkler and drip treatments were not planted. The drip treatments were more frequently irrigated (three times a week with shorter irrigation events) than the drip treatments in the fall trial.

True 6-6-2 organic fertilizer was applied before planting at a rate of 78 kg of N per hectare, and the crop was supplemented with True 4-1-3 liquid fertilizer injection into irrigation system. For the drip system, True 4-1-3 was applied four times after germination (by crop harvest) at a rate of 34, 45, 33, and 44 kg of N per hectare. In the sprinkler system, the same fertilizer was applied at a rate of 45, 45, 50, and 45 kg of N per hectare.

2.2. Field Measurements and Analysis

The actual crop ET (evapotranspiration) was measured using Tule Technology sensor (www. tuletechnologis.com) which uses the residual of energy balance method as surface renewal equipment. Using the actual crop water-use data measured and spatial California Irrigation Management Information System (CIMIS) data (http://wwwcimis.water.ca.gov/SpatialData.aspx), the actual crop coefficient curve was developed for each crop season. The reference ET (ET_o) was retrieved from spatial CIMIS as well. Images were taken on weekly basis utilizing an infrared camera (NDVI digital camera, NDVI stands for the Normalized Difference Vegetation Index) to quantify the development of the crop canopy of each treatment over the crop seasons. The images were analyzed using PixelWrench2 software. Decagon 5TE sensors were installed at three depths (20, 30, and 45 cm) to monitor soil water content on a continuous basis. The numerous horizontal roots of spinach typically remain in the top 30 cm of soil with a spread of approximately 40 cm, while soil water storage at the top 45 cm (spinach crop root zone) was calculated using the soil water content data. The applied water for the irrigation treatments was measured throughout the crop seasons using magnetic flowmeters. The NDVI values and plant leaf wetness values were measured using Spectral Reflectance sensors (combination of SRS-Pi Hemispherical Sensor and SRS-Pr Field Stop Sensor) and dielectric leaf wetness sensors (PHYTOS 31), respectively (METER Group, Inc. USA) on a continuous basis. Leaf chlorophyll was measured using an *atLEAF CHL STD* sensor (FT Green LLC: Wilmington, DE, USA) on a weekly basis.

Shoot biomass (sum of the weight of leaves and stem represents shoot biomass) measurements at the final harvests were carried out in three sample areas of 0.56 m^2 (0.92 m × 0.61 m) per replicate and treatment. The bed located in the center of each replication in each of the treatments was selected as the sample bed (four sample beds per each treatment, a total of 20 sample beds for five irrigation treatments in the fall trial and 14 sample beds for three irrigation treatments in the winter trial). Fresh weight was measured in order to determine shoot biomass accumulation. Leaf chlorophyll content was measured on 20 individual leaves in each of the sample beds. Weekly plant samples (shoots) were analyzed for total nitrogen at the UC Davis Analytical Lab. The statistical significances were performed using generalized linear mixed model using the GLIMMIX (generalized linear mixed models) procedure in SAS 9.4. (SAS Institute Inc., Cary, NC, USA)

2.3. Downy Mildew

Both replicate runs of the experiment were visually scouted by walking down arbitrarily selected rows of all treatments. When disease symptoms were observed, the number of plants exhibiting downy mildew symptoms was counted for the entire length of two of the three beds in each plot. To calculate disease incidence, or the percentage of plants affected by downy mildew, the number of plants in each bed was divided by the estimated plant population. The estimated plant population was determined from the seeding rate and the germination rate averaged over each treatment at emergence. Downy mildew incidence was analyzed in a generalized linear mixed model using the GLIMMIX procedure in SAS 9.4. A block was treated as a random effect, and options in the model statement included use of the beta distribution and the logit link function. Due to evidence for a significant effect of treatment, means were separated using the *lsmeans* statement with the Tukey-Kramer adjustment for multiple comparisons.

3. Results and Discussion

3.1. Weather Conditions

The daily air temperature, relative humidity, and wind speed variations were different over the fall and the winter experiments (Figure 3). At the fall trial, we observed a mean daily air temperature of 18.8 °C at the early- and mid-seasons when the temperature stayed above 6.5 °C at nighttime. The mean daily temperature decreased to 12.1 °C during the late season, and relative humidity dramatically increased over the last 10 days before the final harvest. An average daily wind speed of 1.5 m s^{-1} was observed during the fall experiment.

While a more variable air temperature was observed at the winter experiment, the mean daily temperature was 12.6 °C at the early- and mid-seasons, and the temperature fell below 1.0 °C for a few nights. Higher daytime relative humidity was measured during the early- and mid-seasons compared to the fall. Although there was not a significant difference between the average wind speed of the crop seasons, several windy days occurred during the winter season (hours with a wind speed of more than 8 m s^{-1}). The cumulative ET$_0$ for the fall and winter crops was 122 and 177 mm, respectively.

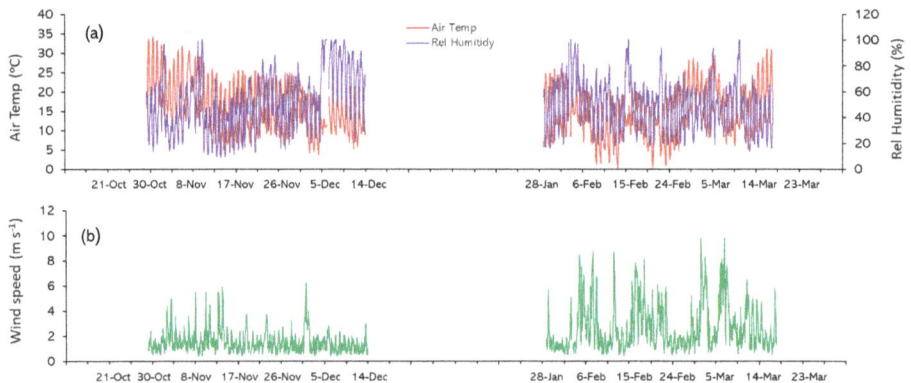

Figure 3. Daily air temperature (**a**), relative humidity (**a**), and wind speed (**b**) over the fall season (October through December, left side of the figures) and the winter season (January through March, right side of the figures).

3.2. Crop Water Use and Applied Water

Total crop water use (seasonal ET) of 92.3 and 154.9 mm was measured in the fall and the winter trials, respectively. Variable daily crop ET and crop coefficient values were measured during the entire crop season (Figure 4). The maximum and minimum crop ET observed was 3.8 and 0.7 mm d^{-1} in the fall experiment and 5.3 and 0.3 mm d^{-1} in the winter experiment, respectively. A similar range of crop coefficient value (0.2–1.2) was obtained for both crop seasons. Piccinni et al. reported a crop water use of 157.5 mm and crop coefficient range of 0.2–1.5 for winter spinach in Texas using lysimeter measurements [20].

The total applied water for the sprinkler treatment was 144.8 and 225.2 mm in the fall and the winter trials, respectively. The total applied water for the drip treatments was 130.5 mm over the fall season and 202.4 mm over the winter season. Overall, an average of 11% more water was applied through the sprinkler system compared with the drip system to compensate for the lower water application uniformity of sprinkler irrigation system. A well-designed and properly managed drip irrigation usually has high distribution uniformity, and unlike conventional sprinkler irrigation systems, has the potential to conserve water because of a lower potential for tailwater runoff and deep percolation losses.

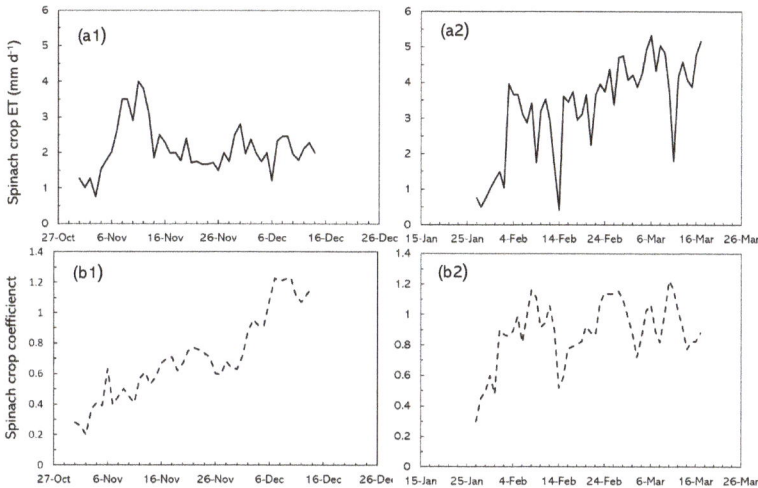

Figure 4. Daily spinach crop evapotranspiration or ET (**a1,a2**) and crop coefficient (**b1,b2**) values in the experimental seasons. The subfigures a1 and a2 show crop ET in the fall and spring trials, respectively, and the subfigures b1 and b2 show crop coefficient values in the fall and spring experimental seasons, respectively.

Soil water storage in the profile was calculated, assuming that the water content sensed at each depth was representative of the soil from that depth to the midpoint between the next upper and lower depths. A comparison of the soil water storage between sprinkler and drip (3.8–4B) treatments versus soil water storage in the average field capacity (FC) of the top 45 cm of the soil is shown in Figure 5. Even though there were differences between the daily available soil water in both treatments and crop seasons, the soil water storage amounts were uniformly maintained at around the average soil water storage of field capacity (19.35 cm at the top 45 cm of the soil). More uniform soil water availability in the drip treatment over the winter experiment could be as a result of more frequent irrigation events with shorter durations.

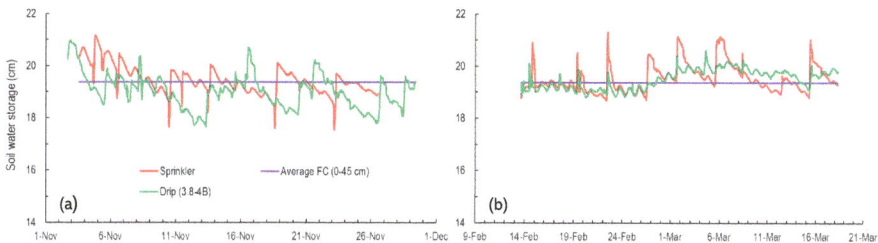

Figure 5. Soil water storage at the surface to a 45 cm-deep profile at the drip (3.8–4B) and sprinkler treatments in the fall (**a**) and winter (**b**) crop seasons. The blue line shows the soil water storage calculated using an average field capacity of the top 45 cm of the soil (0.43 m^3 m^{-3}).

3.3. Crop Canopy over the Season

Crop canopy cover is defined as the percentage of plant material which covers the soil surface, and can be a very useful index for estimating the crop coefficient. Here, the canopy cover percentage was developed for each of the irrigation treatments (Figure 6). Canopy cover percentages show that the leaf density of drip irrigation treatments was slightly behind (1–4 days depending upon the irrigation treatment and crop season) than that of the sprinkler irrigation treatments.

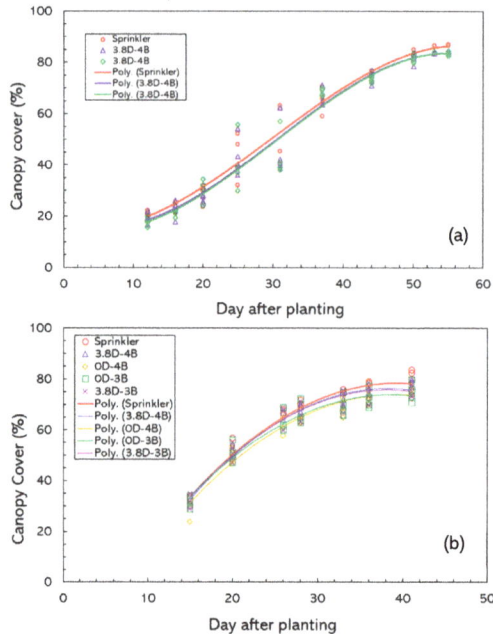

Figure 6. Canopy crop curve for the different irrigation treatments in the (**a**) winter crop season and (**b**) the fall crop season.

The individual canopy cover curves for each season demonstrate that spinach crop water requirements and irrigation scheduling could be different in fall and winter seasons. For instance, in this study, an average of 52% and 70% of canopy crop coverage was observed 30 days after planting in the winter and the fall crop season, respectively. We may expect a longer season for spinach planted in the winter than the fall in the Imperial Valley, though it may change depending on the specific weather conditions. Figure 6 shows the canopy cover for individual experimental beds with four driplines planted at two different dates.

3.4. Crop Growth and Greenness

Few differences between drip and sprinkler treatments were visible for the winter planted trial, while more differences were observed in the fall experiment in November. Leaves began to yellow in between the drip laterals in plots with the three-dripline treatments. A possible reason for this may be that the fertigation did not move the N in between the driplines. The total plant tissue N content of the drip treatments with three driplines in bed was less than the sprinkler treatment on the 30th of November (1.8% N for the drip vs. 3.2% N for the sprinkler), while the difference in N content of the tissue was less between the sprinkler treatment and the drip treatment with four driplines in the bed (Figure 7). Overall, a higher plant-tissue nitrogen content was observed for the sprinkler treatment compared to drip treatments.

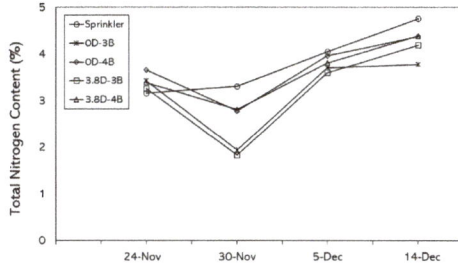

Figure 7. The trends of total plant nitrogen content over the fall crop season. The results are presented for the individual irrigation treatments.

The leaf chlorophyll content (Figure 8) was also higher in the sprinkler treatments compared to the drip treatments, though the drip treatment with four driplines at 3.8 cm depth was numerically more similar to sprinkler treatments and sometime had a greater leaf chlorophyll content during the crop seasons. A greater level of leaf chlorophyll content in the late season was observed at the winter experiment than the fall experiment for both sprinkler and drip treatments. For instance, the total leaf chlorophyll content of the sprinkler treatment two days before the final harvest at the winter trial was 4 µg cm^{-2} more than leaf chlorophyll content of sprinkler treatment at the fall trial. Variable leaf chlorophyll content was observed over the plant development in each of the treatments, which corresponds to the plant N accumulation over the growing seasons.

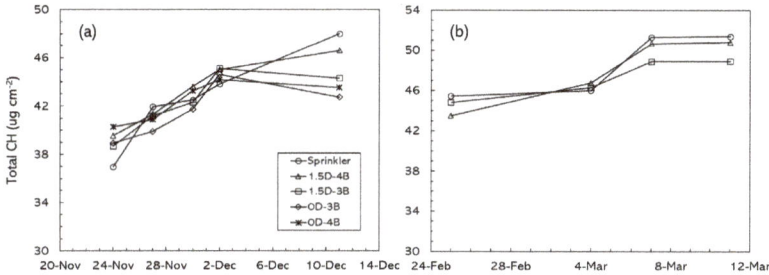

Figure 8. The trends of average leaf chlorophyll content of spinach at different irrigation treatments over (**a**) the fall crop season and (**b**) winter crop season.

Figure 9 shows the average day NDVI values for the drip treatment with four driplines (the 3.8D-4B treatment) and the sprinkler treatment for the fall season experiment. The results indicated that the NDVI values varied from 0.2 (two weeks after planting) to 0.84 (the day before final harvest). By late November, the NDVI values were very similar in both irrigation treatments, where this index had higher values in the drip treatment even over a short period. The results demonstrated lower NDVI values in the drip treatment than sprinkler treatment over the last two weeks of the crop season.

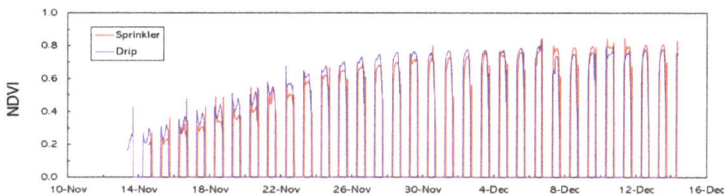

Figure 9. NDVI values of sprinkler and drip (3.8D-4B) treatments over the fall crop season. The daily values vary between a maximum during daytime and zero during nighttime.

The values of total plant nitrogen content, leaf chlorophyll content of spinach, and NDVI confirmed that N uptake at the drip treatments was not entirely as effective as the sprinkler treatment, particularly in the fall experiment. The nutrient management issue in spinach drip irrigation in combination with water management is likely a critical issue that we need to learn more about, since it may affect the adoption and viability of the drip for spinach production. Spinach is a very fast short-season crop, and hence, water and nitrogen management may significantly influence the leaf as the major organ for important physiological processes and essential indicator used to measure the growth and yield of the crop. While the leaf area and shoot biomass of spinach are greatly affected by water and nitrogen levels [4], this dependency is likely more critical in organic spinach.

3.5. Leaf Wetness

Figure 10 shows the probe output of the leaf wetness sensors placed in the sprinkler treatment and drip treatment (3.8D-4B) for a period of 12 days during the fall season experiment. At this period, there were two irrigation events in each of the treatments, and two rainy days. The sensor output from dew was typically lower (less than 700 counts) than that from the rain or the irrigation event. The results revealed that sprinkler-irrigated crop canopies remained wet 24.3% (= (70 h/288 h) × 100) times longer during this period than the crop canopy irrigated with drip treatment.

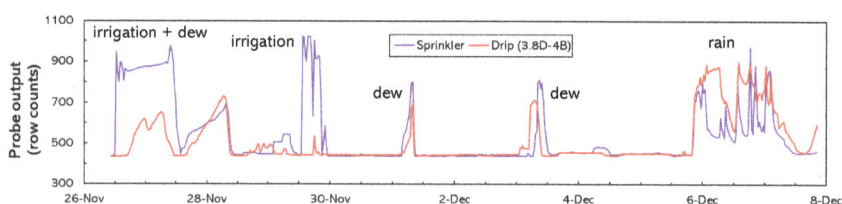

Figure 10. The row counts of leaf wetness sensors at the sprinkler and drip (3.8D-4B) treatments over a 12-day period in the fall crop season.

Spinach downy mildew requires a cool environment with long periods of leaf wetness or high humidity. Wet foliage is especially favorable. Considering the above analysis, and in the case where the weather and farming conditions are similar, there is higher risk for infection and downy mildew disease development in spinach irrigated by sprinklers in comparison with spinach irrigated by drips. The air temperature and relative humidity pattern (Figure 2) indicate that there was a desirable weather condition and more possibility for downy mildew disease in mid-December, but the fall experiment was just terminated at the time. The next desirable period was mid-February when temperatures had cooled to the range believed to be optimal for downy mildew, days became windier, and there was a period of leaf surface wetness caused by sprinkler irrigation, rainfall, or high relative humidity.

3.6. Shoot Biomass

The effects of various irrigation treatments on spinach shoot biomass yield over the two experimental seasons are summarized in Table 2 and Figure 11. In the fall trial, the mean biomass yield in the sprinkler treatment was 13,905 kg ha^{-1}, approximately 9% more than the 3.8D-4B treatment. The lowest mean yield (11,136 kg ha^{-1}) was observed in the 0D-3B treatment. Statistical analysis indicated very strong evidence ($p = 0.001$) for an overall effect of the irrigation system on spinach yield. A significant difference between the individual treatments was investigated using the Tukey-HSD analysis. The results demonstrated a significant yield difference between the sprinkler irrigation and each of the drip irrigation treatments (p values of 0.0001 to 0.0009). Even though no significant yield difference was obtained between the surface drip and sub-surface drip (driplines at 3.8 cm depth) with the same dripline number in bed (p value of 0.8276 for the three-dripline and 0.1995 for the four-dripline), the number of driplines in bed had a very significant impact on spinach biomass yield (*p* values of 0.0001 to 0.0009).

In the winter trial, the mean biomass yield in the sprinkler treatment was 14,886 kg ha^{-1}, approximately 7% more than the 3.8D-4B treatment. Statistical analysis indicated strong evidence (*p* = 0.0424) for an overall effect of irrigation system on spinach fresh yield. While we could not find a significant difference between the impact of the sprinkler and the 3.8D-4B irrigation treatments on spinach yield (*p* = 0.1161), there was a statistically significant yield difference between the sprinkler and the 3.8D-3B irrigation treatments (*p* = 0.04147).

Table 2. Mean spinach fresh yield values of each irrigation treatment in each of the fall and winter experiments. Yields with different letters significantly differ (*p* < 0.05) by Tukey's test.

Fall 2018		Winter 2019	
Irrigation Treatment	**Fresh Yield (kg ha^{-1})**	**Irrigation Treatment**	**Fresh Yield (kg ha^{-1})**
Sprinkler	13,905 [a]	Sprinkler	14,886 [a]
3.8D-4B	12,753 [b]	3.8D-4B	13,914 [ab]
0D-4B	12,273 [b]	3.8D-3B	13,580 [b]
0D-3B	11,136 [c]	-	-
3.8D-3B	11,351 [c]	-	-

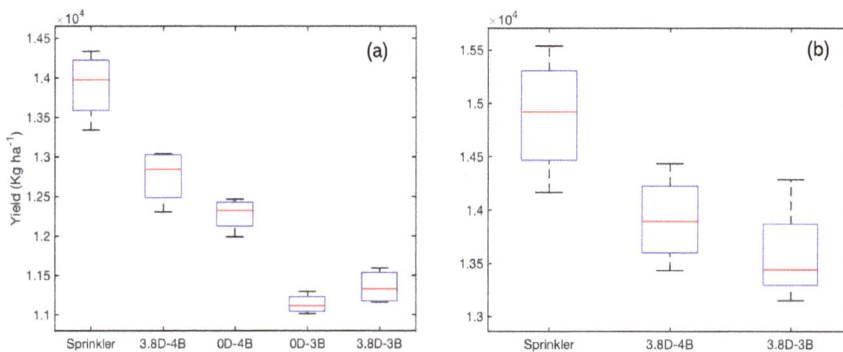

Figure 11. Mean yield data of the sample beds in each of the irrigation treatments at the fall experiment and (**a**) and the winter experiment (**b**). Horizontal red lines indicate the average for each treatment.

The yield reduction in drip irrigation treatments compared to the sprinkler irrigation ranged between 7% (the 3.8D-4B treatment against the sprinkler treatment in the winter trial) and 24.8% (the 0D-3B treatment against the sprinkler treatment in the fall trial). The yield difference may have likely been caused by suboptimal irrigation and nutrient management conditions of the drip treatments. Since drip irrigation was tested for the first time for spinach in this study, subsequent trials need to plan for irrigation and nutrient improvements and be conducted in different aspects. These practices had to be adjusted in real time as the study progressed. The biomass yield reported here are related to a final harvest, which was on the same day for all the treatments. Since the findings showed that the leaf density of drip irrigation treatments was more behind than sprinkler treatment, scheduling a different harvest time for the drip treatments may compensate for some of the yield reductions against the sprinkler treatment.

Several factors influence appropriate drip irrigation management, including system design, soil characteristics, and environmental conditions. The influences of these factors can be integrated into a practical and efficient system which determine the quantity and timing of drip irrigation. Drip irrigation offers the potential for precise water management, and also provides the ideal vehicle to deliver nutrients in a timely and efficient manner. However, achieving high water and nutrient use efficiency while maximizing crop productivity requires intensive and proper management.

However, the 7% through 13% shoot biomass difference between the four dripline treatments and the sprinkler treatment demonstrates the potential of sub-surface drip irrigation for profitable spinach production. This yield difference could be reduced through optimal system design and better irrigation and nutrient management practices for the drip system.

Optimizing water and nitrogen management had not been an objective for this study, therefore, evaluating water use efficiency of the treatments may not be an interest to be discussed. With the 7% lower biomass yield and 11% less applied water in the 3.8D-4B than the sprinkler treatment, we may simply conclude that drip irrigation may have the potential to enhance the efficiency of water use in spinach production or at least at this point, there is no major difference between the two systems regarding this indicator. High water use efficiency for sub-surface drip irrigation was reported by researchers for multiple crops [12,13,15,21].

3.7. Downy Mildew Incidence

Downy mildew was not observed in the fall experiment. In the winter experiment, downy mildew activity was first confirmed in the study area on 5 March. Disease incidence was rated on 11 March. Downy mildew incidence was low on this date, with only two beds (0.12% and 0.20%) exhibiting incidence values above 0.1% (Figure 12). Mean downy mildew incidence in plots irrigated with sprinklers following emergence was 0.08%, approximately 4 to 5× higher than treatments irrigated with drip following emergence. Statistical analysis indicated evidence ($p = 0.0461$) for an overall effect of irrigation treatment on downy mildew.

Analysis of means suggested that all three treatments were statistically similar. However, a pairwise comparison revealed some evidence that downy mildew incidence was lower in plots irrigated with 3.8D-4B ($p = 0.0671$) or 3.8D-3B ($p = 0.1139$) following emergence when compared to the sprinkler.

The likely mechanism causing this effect was a reduction under drip irrigation of leaf wetness, which is critical for infection and sporulation by the downy mildew pathogen. Additional repetitions of this experiment in higher disease pressure situations are needed for further evaluation of the ability of drip irrigation to reduce downy mildew. Another mechanism that could partially account for the observed differences among the treatments is that the leaf density in drip-irrigated plots was slightly behind that of the sprinkler irrigated plots in time. A less dense canopy could reduce the leaf wetness potential, and in turn, disease incidence potential. However, it is unclear if the magnitude of differences in density could account for the magnitude in differences in downy mildew incidence between sprinkler- and drip-irrigated treatments.

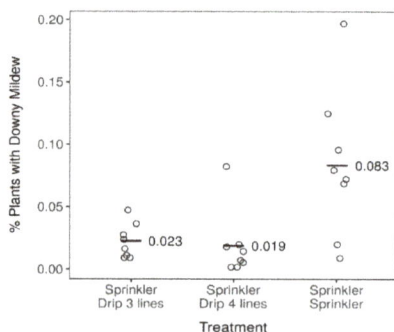

Figure 12. Raw data (all plots and beds within plots) of downy mildew incidence. Horizontal bars and labels indicate the mean for each treatment. Points are jittered to reduce overlap. The top line of *x*-axis labels indicate what the plot was germinated with, and the bottom row indicates the irrigation used following emergence.

Perhaps the most important benefit of drip for spinach production could be less yield loss as a result of downy mildew management. Spinach is a high-value crop—for instance, the value of spinach per kg was about $2 in 2016 [22]. While more data are needed to conduct an economic feasibility analysis of drip irrigation for spinach production, the results of this study demonstrated a positive impression. The initial cost associated with a drip system for spinach fields is estimated at about $3000 to $4000 per hectare in the region.

4. Conclusions

This study demonstrated the potential for drip irrigation to be used to produce organic spinach, conserve water, enhance the efficiency of water use, and reduce incidence of downy mildew. Further work is needed to comprehensively evaluate the viability of utilizing drips—specifically, the optimal system design, the impacts of irrigation and nitrogen management practices in various soil types and climates, and strategies to maintain productivity and economic viability at spinach. Assessing drip irrigation for the entire crop season, including germination, could be another research interest since spinach is a short-season crop and combining the sprinkler for crop germination and drip for such a short period might cause some practical issues.

Author Contributions: Conceptualization, A.M. and M.C.; Data curation, A.M.; Formal analysis, A.M.; Investigation, A.M. and A.P.; Methodology, A.M., M.C. and A.P.; Project administration, A.M.; Supervision, A.M.; Writing—original draft, A.M.; Writing—review & editing, A.M., M.C. and A.P.

Funding: This research was funded by the California Leafy Greens Research Board.

Conflicts of Interest: The authors declare no conflict of interest.

References

1. California Department of Food and Agriculture (CDFA). *California Agricultural Statistics Review*; Agricultural Statistical Overview: Sacramento, CA, USA, 2018.
2. Ortiz, C. *Office of the Agricultural Commissioner Sealer of Weights and Measures-Imperial County*; Annual Crop and Livestock Report: Imperial County, CA, USA, 2018.
3. Birdsall, S. *Office of the Agricultural Commissioner Sealer of Weights and Measures-Imperial County*; Annual Crop and Livestock Report: Imperial County, CA, USA, 2009.
4. Zhang, J.; Sha, Z.; Zhang, Y.; Bei, Z.; Cao, L. The effects of different water and nitrogen levels on yield, water and nitrogen utilization efficiencies of spinach. *Can. J. Plant Sci.* **2015**, *95*, 671–679. [CrossRef]
5. Cantliffe, D.J. Nitrate accumulation in vegetable crops as affected by photoperiod and light duration. *J. Amer. Soc. Hort. Sci.* **1972**, *97*, 414–418.
6. Heinrich, A.; Smith, R.; Cahn, M. Nutrient and Water Use of Fresh Market Spinach. *HortTechnology* **2013**, *23*, 325–333. [CrossRef]
7. Zhang, J.; Bei, Z.; Zhang, Y.; Cao, L. Growth Characteristics, Water and Nitrogen Use Efficiencies of Spinach in Different Water and Nitrogen Levels. *Sains Malays.* **2014**, *43*, 1665–1671.
8. Choudhury, R.A.; Koike, S.T.; Fox, A.D.; Anchieta, A.; Subbarao, K.V.; Klosterman, S.J.; McRoberts, N. Season-Long Dynamics of Spinach Downy Mildew Determined by Spore Trapping and Disease Incidence. *Phytopathology* **2016**, *106*, 1311–1318. [CrossRef] [PubMed]
9. Correll, J.C.; Bluhm, B.H.; Feng, C.; Lamour, K.; Du Toit, L.J.; Koike, S.T. Spinach: Better management of downy mildew and white rust through genomics. *Eur. J. Plant Pathol.* **2011**, *129*, 193–205. [CrossRef]
10. Correll, J.C.; Feng, C.; Matheron, M.E.; Porchas, M.; Koike, S.T. Evaluation of spinach varieties for downy mildew resistance. *Plant Dis. Manag. Rep.* **2017**, *11*, 122.
11. Koike, S.; Cahn, M.; Cantwell, M.; Fennimore, S.; Natwick, E.; Smit, R.; Takele, E. *Spinach Production in California*; University of California Agriculture and Natural Resources: Auckland, CA, USA, 2011; ISBN 978-1-60107-719-6.
12. Montazar, A.; Bali, K.; Zaccaria, D.; Putnam, D. Viability of subsurface drip irrigation for alfalfa production in the low desert of California. In Proceedings of the ASABE International Meeting, Detroit, MI, USA, 29 July–1 August 2018.

13. Hartz, T.K.; Bottoms, T.G. Nitrogen requirements of drip-irrigated processing tomatoes. *HortScience* **2009**, *44*, 1988–1993. [CrossRef]
14. Lamm, F.R.; Camp, C.R. Subsurface drip irrigation. In *Microirrigation for Crop Production—Design, Operation and Management*; Lamm, F.R., Ayars, J.E., Nakayama, F.S., Eds.; Elsevier Publications: Amsterdam, The Netherlands, 2017; Chapter 13; pp. 473–551.
15. Lamm, F.R.; Bordovsky, J.P.; Schwankl, L.J.; Grabow, G.L.; Enciso-Medina, J.; Peters, R.T.; Colaizzi, P.D.; Trooien, T.P.; Porter, D.O. Subsurface Drip Irrigation: Status of the Technology. In *2010 5th National Decennial Irrigation Conference*; Phoenix Convention Center: Phoenix, AZ, USA, 5–8 December 2010.
16. Montazar, A.; Putnam, D.; Bali, K.; Zaccaria, D. A model to assess the economic viability of alfalfa production under subsurface Drip Irrigation in California. *Irrig. Drain.* **2017**, *66*, 90–102. [CrossRef]
17. Luque Quispe, M.R. Evaluation of drip and Microasperation Irrigation Methods in Spinach (Spinaca oleracea) and Swiss Lettuce (Valerianella locusta) in Walipinis. Brigham Young University. Available online: https://scholarsarchive.byu.edu/etd (accessed on 13 June 2019).
18. Takebe, M.; Okazaki, K.; Kagishita, K.; Karasawa, T. Effect of drip fertigation system on nitrate control in spinach *Spinacea oleracea L.. Soil Sci. Plant Nutr.* **2006**, *52*, 251. [CrossRef]
19. ASAE Standards. *Procedure for Sprinkler Testing and Performance Reporting*; ASAE: St. Joseph, MI, USA, 1985.
20. Piccinni, G.; Ko, J.; Marek, T.; Leskovar, D.I. Crop Coefficients Specific to Multiple Phenological Stages for Evapotranspiration-based Irrigation Management of Onion and Spinach. *HotrtScience* **2009**, *44*, 421–425. [CrossRef]
21. Montazar, A.; Radawich, J.; Zaccaria, D.; Bali, K.; Putnam, D. Increasing water use efficiency in alfalfa production through deficit irrigation strategies under subsurface drip irrigation. Water shortage and drought: From challenges to solution. In Proceedings of the USCID Water Management Conference, San Diego, CA, USA, 17–20 May 2016.
22. Ortiz, C. *Office of the Agricultural Commissioner Sealer of Weights and Measures–Imperial County*; Annual Crop and Livestock Report: Imperial County, CA, USA, 2017.

agriculture

MDPI

Article

Deficit Drip Irrigation in Processing Tomato Production in the Mediterranean Basin. A Data Analysis for Italy

Rosa Francaviglia and Claudia Di Bene *

Council for Agricultural Research and Economics, Research Centre for Agriculture and Environment (CREA-AA), 00184 Rome, Italy; rosa.francaviglia@crea.gov.it
* Correspondence: claudia.dibene@crea.gov.it

Received: 8 April 2019; Accepted: 17 April 2019; Published: 19 April 2019

Abstract: In this study, the effects of deficit irrigation (DI) on crop yields and irrigation water utilization efficiency (IWUE) of processing tomato are contrasting. This study aimed at analyzing a set of field experiments with drip irrigation available for Mediterranean Italy in terms of marketable yields and IWUE under DI. Both yields and IWUE were compared with the control treatment under full irrigation, receiving the maximum water restoration (MWR) in each experiment. The study also aimed at testing the effect of climate (aridity index) and soil parameters (texture). Main results indicated that yields would marginally decrease at 70–80% of MWR and variable irrigation regimes during the crop cycle resulted in higher crop yields. However, results were quite variable and site-dependent. In fact, DI proved more effective in fine textured soils and semiarid climates. We recommend that further research should address variable irrigation regimes and soil and climate conditions that proved more unfavorable in terms of crop response to DI.

Keywords: deficit irrigation; Mediterranean region; tomato fruit yield; irrigation water use efficiency

---

1. Introduction

Water resources are extremely scarce in many areas of the world, and water saving has become a priority due to the increase in population and global climate change [1,2]. Agriculture is a major water consumer in regions where irrigation is required for profitable yields, and strategies to reduce water use have the potential to increase sustainability of production. Globally, agricultural irrigation is responsible for 70–80% of freshwater consumption [3,4]. Increased water savings and optimization of irrigation management as much as possible are, thus, urgently needed. A small amount of water saved can be used for other purposes. Therefore, in recent decades, agricultural water use efficiency has been improved by innovations in technology and plant breeding. Irrigation systems and scheduling mainly affect crop yields. Thus, the knowledge of crop water requirement, the reference crop evapotranspiration, and the rainfall of the target region is recommended [5]. While full irrigation (FI) aims to meet crop water requirements to maximize crop yield, in deficit irrigation (DI) water use is optimized in relation to crop yield per volume of water consumed. Modest yield reductions can be acceptable if connected to a significant reduction in water use [6]. DI has been found suitable for grapevine and fruit crops, but vegetables might suffer from losses in yield and quality. DI effects on crop yield and water efficiency have been studied on several crops, including tomato (*Solanum lycopersicum* L.), though with contrasting results that can be ascribed to the different cultivars cultivated or the period of DI application during the crop cycle [6]. Although farm income is higher with increasing yields when more water is supplied with irrigation, water availability is continuously decreasing due to the competing requirements of agriculture, industry, recreation, and the environment. In addition, DI provides an effective adaptive response to water scarcity within a climate change perspective [7], and lower yields might be compensated by the increased production and farm income from additional lands irrigated with the water saved by DI [8,9].

Most of horticultural production areas are in hot and dry climates due to favorable weather conditions (high light, high temperature), as in Mediterranean regions, however, soil water deficit is rather frequent. Tomato is considered one of the most commonly consumed vegetables and economically important crops in the world, and has the highest planted areas of all vegetables worldwide. It is characterized by high-water needs [10] due to the high temperatures and the large gap between rainfall and evapotranspiration (ET) during the long spring-summer growing season [2]. Tomato is a drought sensitive plant because its yield decreases considerably after short periods of water deficiency [5]. The application of DI strategies during tomato growing season may greatly contribute to saving irrigation water [11,12] without affecting tomato yield, compared with (FI) receiving the maximum water restoration (MWR), at a rate of 100% ET [13]. Nevertheless, water deficits at different growth stages can differentially affect tomato yield. Results of crop simulation models showed that certain tomato life stages, such as the flowering and fruiting stages, were more susceptible to water stress than the seedling stage [14].

According to the World Processing Tomato Council [15] (an international non-profit organization representing the tomato processing industry), Italy is ranked as the second producer of processing tomato worldwide, after California and followed by China. Italy is the leading country in Europe, contributing to 44% of the total amount, followed by Spain (27%), Turkey (12%) and Portugal (11%). The total national production in 2018 was 4,811,955 t, cultivated on a surface of 72,504 ha (average marketable yield is 66 t ha^{-1}). The two most important production regions are Emilia-Romagna in the north and Apulia in the south (concentrated in the Capitanata plain, in Foggia province). The two regions contributed in 2018 to 35 and 32% respectively of the national production of processing tomato, as reported by official statistics [16]. Average yields were 69 and 85 t ha^{-1}, respectively. In these areas processing tomato cultivation is highly intensive due to large and regular application of irrigation water and nutrient inputs during flowering and fruit formation [17,18], which may create the potential for negative side-effects on the environment [19]. Thus, the application of water saving strategies is of particular interest where water availability is limited and to save water while maximizing tomato yields under water deficit conditions [20,21]. Since the price of water is increasing, DI is an effective strategy to provide an adequate economic profit for farmers in Mediterranean environments [22]. Moreover, results of crop simulation models in southern Italy have shown that climate change would decrease tomato yields due to the shorter crop cycle induced by the temperature increase [23,24]. Besides water savings, gains in fruit quality (higher soluble solid contents and fruit color intensity) can often compensate for the losses in fruit yields [11,21]. However, the contrasting results available in the scientific literature suggest the need to better understand site-dependent plant responses to water deficit with DI [25,26].

A quantitative analysis is important to provide suggestions for improving crop yield and recommendations of irrigation water inputs in processing tomato cultivation under Mediterranean conditions. The present study is aimed at evaluating the effect of DI irrigation on processing tomato in field experiments derived from a literature search. Data were analyzed in terms of marketable yields, water restoration, and irrigation water use efficiency (IWUE) under DI compared with the control treatment under full irrigation, receiving the MWR in each experiment. The study also aimed at testing the effect of climate (aridity index) and soil parameters (texture).

2. Materials and Methods

2.1. Data Collection and Case Studies

To assess the effect of deficit drip irrigation on tomato yields, a data search of existing field studies was performed. The literature search was performed with SCOPUS with no source limitations (all years, article types, and access types). Literature was screened by searching three fields in the title, abstract, and keywords of the source reference: "Mediterranean" AND "Italy" AND "tomato". Results referring to greenhouse studies, pot experiments, Life Cycle Assessment, and simulation studies addressing crop development and water dynamics were excluded from the data analysis.

Information derived from field experiments included: region, province, altitude (m above sea level), long-term mean annual temperature (MAT °C) and total rainfall (MAP mm), aridity index, rainfall and irrigation during the growth cycle, marketable yield as fresh and dry matter, irrigation treatments, and soil texture group (Table S1). Crop evapotranspiration (ETc) in the different field experiments was estimated by the authors as the product of reference evapotranspiration (ETo), calculated with the FAO Penman–Monteith equation [27] or using Class A pan evaporation and Kpan [28], and tomato-specific crop coefficient (Kc). All experiments preformed were under drip irrigation.

Using these criteria 10 studies, totaling 54 yield observations, were found in four regions of Italy: Apulia (3), Basilicata (3), Latium (2), and Sicily (2).

2.1.1. Apulia

A two-year field research (2011–2012) was carried out at Valenzano (41°03′N, 16°52′E, altitude 72 m a.s.l) in Bari Province [29]. MAT and MAP were 16.2°C and 523 mm, respectively. Tomato (cv. Tomato F1) was grown under three irrigation regimes: full recover of crop evapotranspiration (I100), 50% of full irrigation supply (I50), and rainfed (I0). Tomato was transplanted in mid-April, and fertilized with 100, 120, and 150 kg ha^{-1} of N, P_2O_5, and K_2O, respectively.

A field research was carried out in 2011 in Foggia province (41°45′N, 15°50′E, altitude 90 m a.s.l.), with MAT and MAP of 15.8°C and 526 mm, respectively [22]. Four irrigation regimes re-establishing 125% (ET125), 100% (ET100), 75% (ET75), and 50% (ET50) of ETc were considered. Tomato (cv. Defender F1) was transplanted in mid-May and fertilized with 133, 75, and 90 kg ha^{-1}, respectively, of N, P_2O_5, and K_2O.

A two-year experiment (2009–2010) was carried out in Foggia province (41°24′N, 15°45′E., altitude 30 m a.s.l) with MAT and MAP of 15.8°C and 526 mm respectively [30]. Tomato was cultivated under four irrigation regimes: DI, constant regime with restoration of 60% of maximum ETc during the crop cycle; RDI, variable irrigation regime with 60%, 80%, and 60% of maximum ETc through the three main phenological stages of the crop cycle; FI, full irrigation regime with the restoration of 100% ETc; FaI, farmer irrigation regime based on usual farming routine. Tomato (cv. Genius F1) was transplanted in the first decade of May and fertilized with 154 and 56 kg ha^{-1} of N and P_2O_5, respectively.

2.1.2. Basilicata

A two-year experiment (2002–2003) was carried out in Lavello (41°03′N, 15°42′E, altitude 180 m a.s.l) in Potenza province [31]. MAT and MAP were 14.5°C and 518 mm, respectively. Tomato was cultivated under six irrigation regimes: (i) four constant irrigation regimes with restoration of 0 (T0), 50 (T1), 75 (T2), and 100% (T3) of ETc during the whole crop cycle; (ii) two variable irrigation regimes with 100% restoration of ETc during the first period of the crop growth, followed by 75 or 50% restoration of ETc in the second part of the cycle (T4 and T5 treatments respectively). Tomato (cv. Pullrex) was transplanted after mid-May, and fertilized with 182, 214, and 160 kg ha^{-1} of N, P_2O_5, and K_2O, respectively.

A first field experiment carried out at Metaponto (40°24′N, 16°48′E, altitude 10 m a.s.l) in Matera province [32] reported the results related to 2007 and 2009 growing cycles. MAT and MAP were 16.5°C and 493 mm, respectively. Three irrigation treatments were compared: re-establishing 50 (I1), 75 (I2), and 100% (I3) of the crop evapotranspiration (ETc). Tomato (cv Tomito) was transplanted in mid-May and fertilized with 180 kg ha^{-1} of N.

A second two-year experiment (2008–2009) in the same area compared three irrigation regimes: V100, full restoration (100%) of ETc, V50, 50% restoration of ETc, and V0, no water restoration [33]. Tomato (cv. Faino F1) was transplanted in late May and fertilized with 160 kg ha^{-1} of N.

2.1.3. Latium

Two field experiments were carried out in Viterbo province (42°43′N, 12°07′E, altitude 310 m a.s.l). MAT and MAP were 14.4°C and 746 mm respectively. The first research [34] was conducted in 1997, and compared four irrigation regimes: 50–75, 50–100, 75–50, and 100–75 % restitution of ETc in the first (from planting to fruit set) and in the second (from fruit set to harvest) growth period. Tomato (hybrid PS 1296) was transplanted at the end of May with three fertilization treatments: control (no fertilization), D1 with 79, 68, and 107 kg ha^{-1} of N, P_2O_5, and K_2O respectively, and D2 (double the doses of D1).

The second experiment [35] was carried out in 2006–2007 and compared two irrigation treatments: full irrigation (FULL) restoring 100% of ETc, and deficit irrigation (DI), restoring 50% of ETc. Tomato (cv. Carioca). which were transplanted in mid-May and fertilized with 152, 200, and 150 kg ha^{-1} of N, P_2O_5, and K_2O, respectively.

2.1.4. Sicily

A two-year field experiment [12] was carried out in 2001–2002 in Enna province (37°27′N, 14°14′E, altitude 550 m a.s.l.). MAT and MAP were 15.4°C and 514 mm respectively. Four irrigation treatments were compared: no irrigation after plant establishment (V0), 100% (V100) or 50% (V50) ETc restoration up to fruit maturity, 100% ETc restoration up to flowering, then 50% ETc restoration (V100-50). Tomato (cv. Brigade) was transplanted in early May and fertilized with 150, 229, and 120 kg ha^{-1} of N, P_2O_5, and K_2O, respectively.

A field experiment [26] was conducted in 2002 in Siracusa province (37°03′N, 15°18′E, altitude 10 m a.s.l.). MAT and MAP were 17.8°C and 504 mm, respectively. Five irrigation regimes were compared: no irrigation after plant establishment (NI), long-season full irrigation with 100% ETc restoration (LF), long-season deficit irrigation with 50% ETc restoration (LD), short-season full irrigation up to first fruit set with 100% ETc restoration (SF), and short-season deficit irrigation up to first fruit set with 50% ETc restoration (SD). Tomato (cv. Brigade) was transplanted in early May and fertilized with 150, 229, and 120 kg ha^{-1} of N, P_2O_5, and K_2O respectively.

2.2. Data Evaluation

Marketable fruit yields (Mg ha^{-1} fresh weight) under deficit irrigation (DI) were compared with yields of the control treatment with the maximum water restoration (MWR) of each experiment, including rainfall:

$$\text{Yield (\%)} = \text{Yield}_{DI}/\text{Yield}_{MWR} \times 100 \qquad (1)$$

Irrigation water use efficiency (IWUE) of the different treatments was calculated according to [36]:

$$\text{IWUE} = \text{Yield/TWS} \qquad (2)$$

where Yield is the fruit dry biomass at harvest (kg ha^{-1}), and TWS is the total water supply including irrigation and rainfall from planting to harvest (m^3 ha^{-1}).

The Aridity index [37] was calculated with the formula Aridity index = MAP/(MAT+10) that defines aridity classes as humid (30–60), sub-humid (20–30), semi-arid (15–20), arid (5–15), and strongly-arid (< 5). Total water supplies with deficit irrigation during the crop cycle were divided in five classes based on the % of maximum water restoration (MWR): 0–20, 20–40, 40–60, 60–80, and 80–100%. Soil texture group was evaluated according to Soil Taxonomy [38] as (C) coarse (sandy loam, sandy clay loam, loamy sand), (M) medium (clay loam, loam, silty clay loam, silt, silt loam), and (F) fine (clay, silt clay, sandy clay).

Statistical analyses were performed using Statistica 7.0 (Statsoft, Tulsa, OK, USA). Significant differences among means were evaluated through the Fisher's protected least significant difference test (LSD post hoc test).

3. Results

A summary of maximum marketable fruit yields (Mg fresh weight ha^{-1}) under full irrigation and the related total water supply (mm) by rainfall and irrigation during the crop cycle in the different provinces are shown in Figures 1 and 2. Marketable yields ranged from 114.2 to 51.0 Mg ha^{-1}, respectively, at Matera and Siracusa. Total water supplies ranged from 768 to 395 mm at Foggia and Enna, respectively. The result for Matera, coupling a high marketable fruit yield (114.2 Mg ha^{-1}) and a low water supply (517 mm), are an indication of a proper irrigation schedule when fully restoring crop evapotranspiration. Conversely, at Foggia a slightly lower marketable fruit yield (95.2 Mg ha^{-1}) was coupled with a higher water supply (768 mm), indicating the ineffectiveness in productive terms of water supplies following the farmer routine, using more water than the full irrigation regime, restoring 100% of ETc [30].

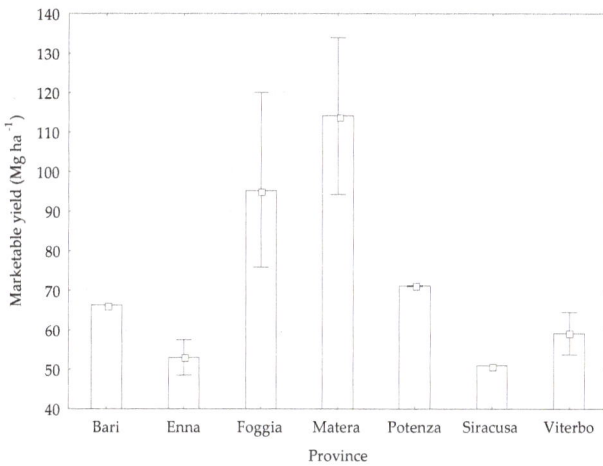

Figure 1. Maximum marketable fruit yield (Mg fresh weight ha^{-1}) in the different provinces under full irrigation. Boxes represent mean values, whiskers represent Min–Max interval.

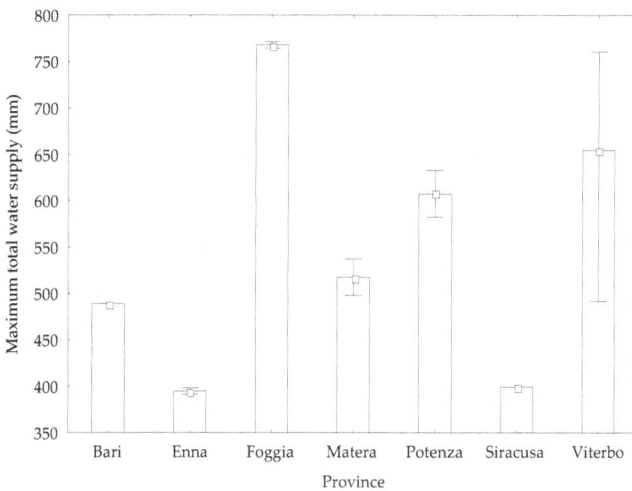

Figure 2. Total water supply (mm) under full irrigation in the different provinces. Boxes represent mean values, whiskers represent Min–Max interval.

3.1. Marketable Yield and Water Restoration

When comparing marketable fruit yields under DI with the control treatment (Equation (1)), average yields (%) significantly differed among irrigation classes (p = 0.0000). In detail (Figure 3), yields were significantly lower in 0–20 and 20–40 irrigation classes (31.5 and 27.3 % respectively) and higher in 40–60, 60–80, and 80–100 classes (74.9, 72.6, and 87.4% respectively).

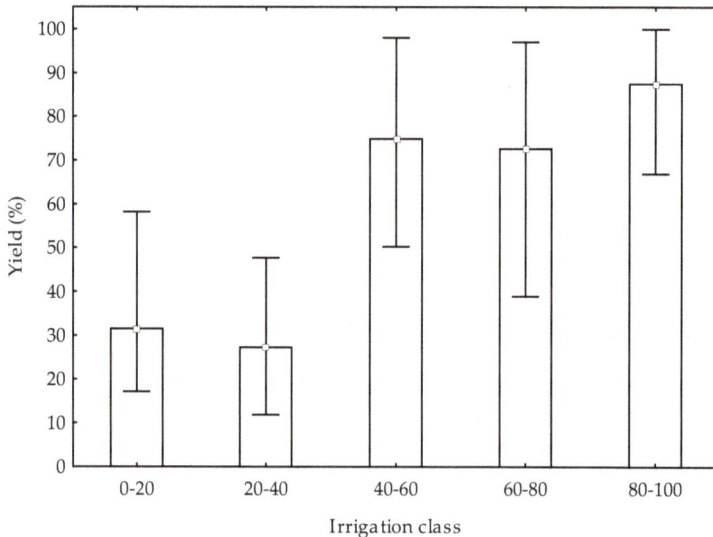

Figure 3. Tomato fresh fruit yield (%) compared with the control treatment based on the irrigation classes as % of maximum water restoration (MWR), F (4;35) = 14.1531; p = 0.00000. Boxes represent mean values, whiskers Min–Max interval.

Average yields (%) significantly differed among irrigation regimes (p = 0.0000). As expected, yields (Figure 4) were significantly lower with none irrigation excluding rainfall during the crop cycle (26.8%), and did not differ between constant and variable regimes with the regulated deficit irrigation (RDI) but were lower under constant irrigation (74.0%) in comparison with RDI (85.7%).

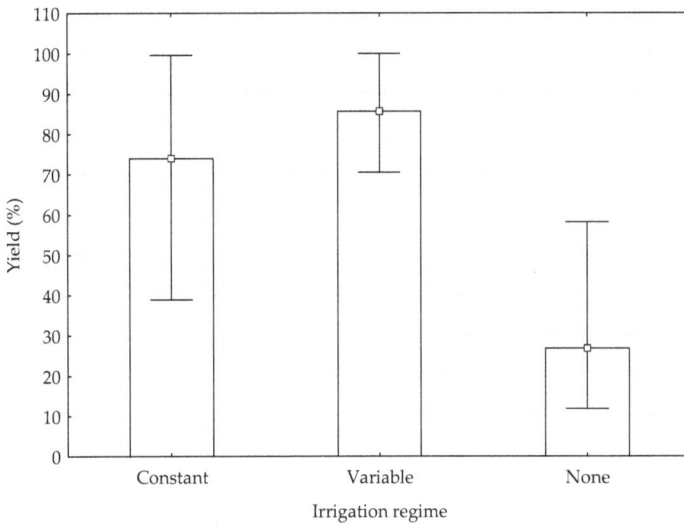

Figure 4. Tomato fresh fruit yield (%) based on the irrigation regimes, F (2;37) = 25.6319; p = 0.00000. Boxes represent mean values, whiskers represent Min–Max interval.

Fresh fruit yield (%) and maximum water restoration (%) supplied with DI were interpolated (Figure 5) with a polynomial equation: $y = -0.0039x^2 + 1.2053x + 16.8326$ ($R^2 = 0.7045$). The interpolating function indicated that yields would decrease by 6.3, 8.9, and 11.7% at 90, 85, and 80% of maximum water restoration (MWR), but would still be acceptable at 75 and 70% of MWR with decreases of 14.7 and 17.9%.

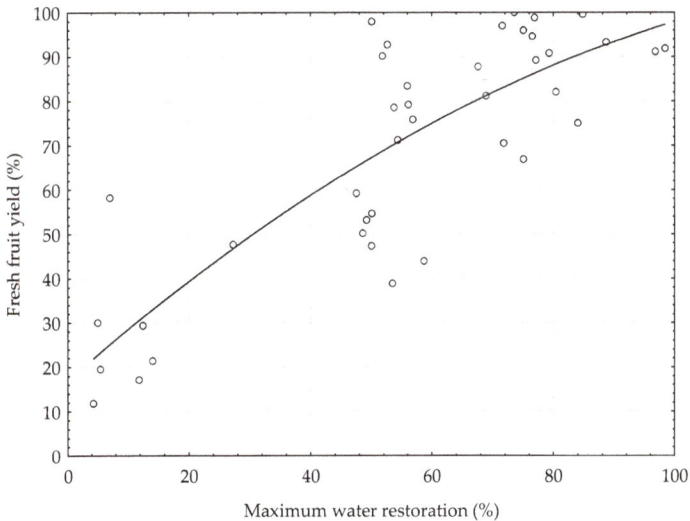

Figure 5. Tomato fresh fruit yield (%) as a function of maximum water restoration (%).

Marketable fruit yields (Mg fresh weight ha^{-1}) under DI significantly differed among the provinces (p = 0.00006) where the field experiments were conducted (Figure 6). Average yields were significantly lower at Bari, Siracusa, Enna, Viterbo, and Potenza (22.2, 27.4, 36.0, 41.1, and 50.3 Mg ha^{-1}, respectively) compared to Foggia and Matera (75.4 and 92.7 Mg ha^{-1}, respectively). Compared with marketable yields under full irrigation (Figure 1), decreases were as follows: Bari (67%), Siracusa (46%), Enna (32%), Viterbo (31%), Potenza (29%), Foggia (21%), and Matera (19%).

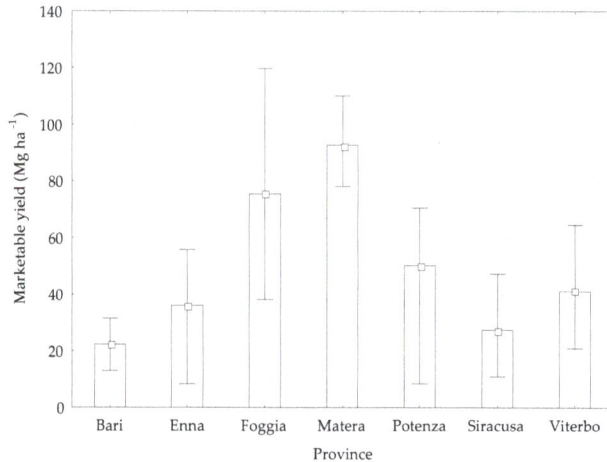

Figure 6. Marketable yield (Mg fresh weight ha^{-1}) in the different provinces under DI, F (6;33) = 7.1298; p = 0.00006. Boxes represent mean values, whiskers represent Min–Max interval.

Marketable fruit yields (Mg fresh weight ha^{-1}) under DI also significantly differed among soil texture groups (p = 0.0049) and average yields (Figure 7) were significantly lower in field experiments with coarse and medium texture (32.5 and 51.7 Mg fresh weight ha^{-1}, respectively) compared to fine textured soils (75.4 Mg fresh weight ha^{-1}). This result is coherent with the soil textures of the field experiments, which were fine and medium at Foggia and Matera, respectively, and also showed the lowest yield decreases; conversely, at Bari soils were coarse textured and presented the highest yield decrease under DI (Figure 6).

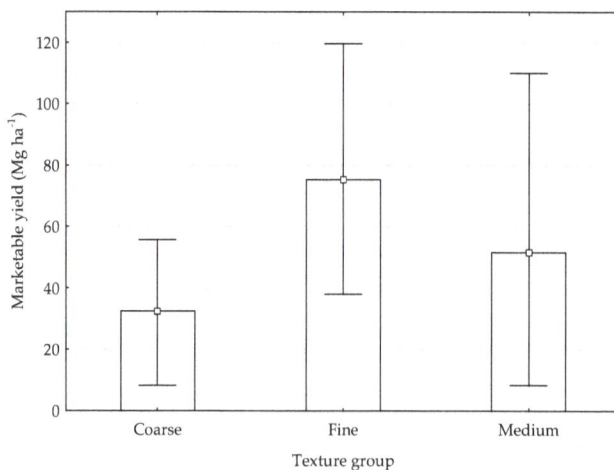

Figure 7. Marketable yield (Mg fresh weight ha^{-1}) under DI based on soil texture groups according to United States Department of Agriculture (USDA), F (2;37) = 6.1615; p = 0.0049. Boxes represent mean values, whiskers represent Min–Max interval.

3.2. Irrigation Water Use Efficiency and Water Restoration

Irrigation water use efficiency (IWUE in kg dry weight m^{-3}) was weakly significantly different ($p = 0.0683$) among irrigation classes (Figure 8). Average IWUE was significantly higher only in 0–20 irrigation class (1.67 kg m^{-3}) compared to the other classes.

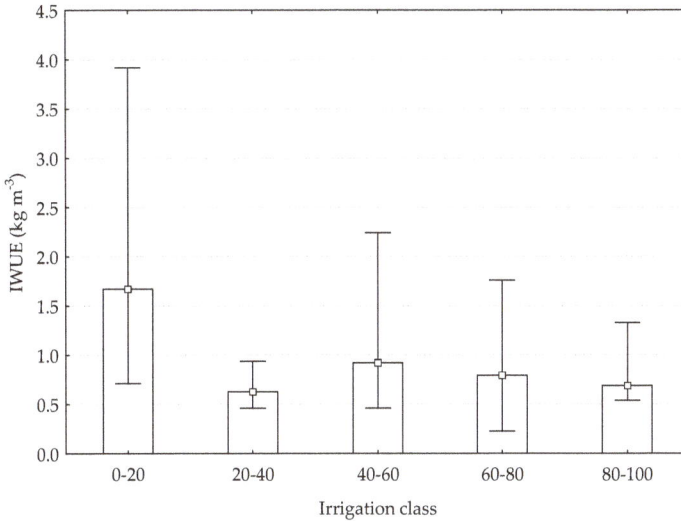

Figure 8. Irrigation Water Use Efficiency (IWUE) (kg dry weight m^{-3}) based on the irrigation classes as % of maximum water restoration (MWR), F (4;35) = 2.403; $p = 0.0683$. Boxes represent mean values, whiskers represent Min–Max interval.

In relation to irrigation regimes (Figure 9), IWUE was not significantly different ($p = 0.2366$). IWUE was higher with no irrigation (1.18 kg m^{-3}) and decreased with constant and variable irrigation regimes (0.86 and 0.66 kg m^{-3}, respectively).

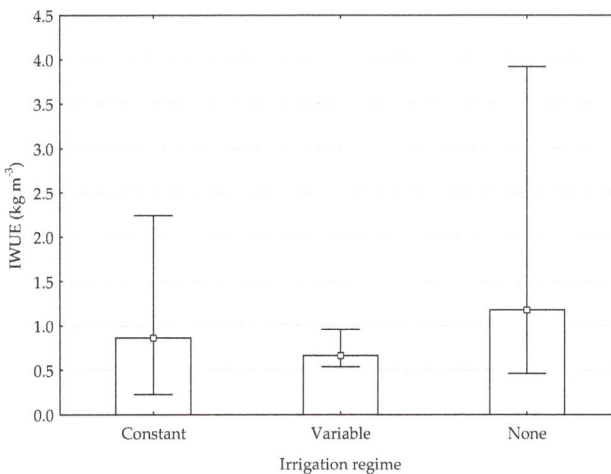

Figure 9. Irrigation Water Use Efficiency (IWUE) (kg dry weight m^{-3}) based on irrigation regimes, F (2;37) = 1.4992; $p = 0.2366$. Boxes represent mean values, whiskers represent Min–Max interval.

IWUE significantly differed among the provinces ($p = 0.0000$) where the field experiments were conducted (Figure 10). In detail, average IWUE was significantly lower at Viterbo, Bari, Potenza, Foggia Siracusa, and Enna (0.47, 0.49, 0.61, 0.75, 0.80, and 0.98 kg m^{-3}, respectively) compared to Matera (2.31 kg m^{-3}). Results for Matera are in agreement with the low decreases observed in marketable yields.

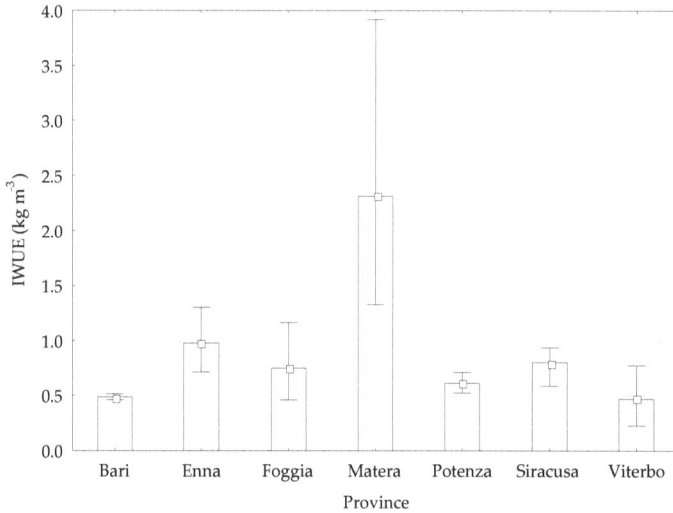

Figure 10. Irrigation Water Use Efficiency (IWUE) (kg dry weight m^{-3}) in the different provinces, F (6;33) = 11.5912; $p = 0.0000$. Boxes represent mean values, whiskers represent Min–Max interval.

IWUE also significantly differed among aridity classes ($p = 0.0092$) and on average was significantly lower (Figure 11) in field experiments under humid and sub-humid climates (0.47 and 0.75 kg m^{-3}, respectively) compared to semiarid conditions (1.34 kg m^{-3}). Considering the location of field experiments, humid climate conditions are related to Viterbo in Latium, and semiarid conditions to Matera in Basilicata.

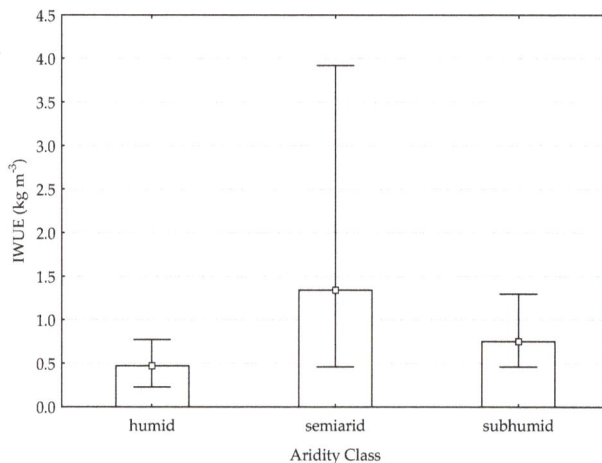

Figure 11. Irrigation Water Use Efficiency (IWUE) (kg dry weight m^{-3}) based on the aridity classes [37], F (2;37) = 5.3391; $p = 0.0092$. Boxes represent mean values, whiskers represent Min–Max interval.

4. Discussion

In the environments typical of the Mediterranean area the use of water resources for irrigation is a priority to be managed through sustainable regulation of water supplies, contributing to water savings with environmental and economic benefits [39] but avoiding high productivity losses to maintain profit for farmers [6,30]. Research also indicated that DI may have positive side effects, such as contributing to decreased soil CO_2 emissions and enhanced C sequestration in soils, by decreasing microbial activity in response to decreased soil moisture levels [39]. In addition, nitrogen fertilization also results in lower N_2O emissions in Mediterranean regions with drip irrigation systems that are commonly used in tomato cultivation compared with sprinkler irrigation methods [40].

Results from the different field experiments examined in this study are in contrast with each other. Data obtained in a study at Foggia in Apulia [30] indicated that farmers tend to over irrigate tomato crops, with no significant increase in the marketable fruit yield and quality, as reported in other research [20]. Moreover, the same authors [30] indicated that the adoption of variable irrigation regimes as RDI restoration of 60%, 80%, and 60% of the maximum ETc during the three main tomato phenological stages (i.e., from plant establishment to flowering of the first truss, from flowering of the first truss to fruit breaking colors of the first truss, and from fruit breaking colors of the first truss to harvest), was effective to save water, as shown by other authors [31,34] at Viterbo (Latium) and Lavello (Basilicata). A study at Matera in Basilicata [32] indicated that water restoration of 50, 75, and 100% of crop evapotranspiration showed no statistical differences among the irrigation volumes in relation to tomato yield and quality. Conversely, in the same environment another study [33] reported statistically significant differences in both marketable yields and fruit quality when restoring 0, 50, and 100% of ETc.

The study conducted at Enna in Sicily [12] showed that marketable yields were strongly decreased by early soil water deficit following plant establishment, while a reduced irrigation rate after the initial stages or after flowering did not induce any significant loss. The study also indicated that DI has beneficial effects on fruit quality. In particular, a high total solids content of the fruit improves the efficiency of the industrial process due to the lower energy required to evaporate water from fruit. Tomato yield also proved more sensitive to the length of the irrigation period rather than to the total water supplied during DI experiments in Sicily [26]. In fact, the long-season deficit irrigation (LD) with 50% ETc restoration and the short-season full irrigation (SF) with 100% ETc restoration received about the same amount of water, but yields decreased by 46% in SF. In addition, irrigation cut-off during the ripening period did not significantly affect marketable yields and enhanced fruit quality [31].

In the case of Foggia and Viterbo, the average irrigation supplied (Figure 2) was the highest (768 and 655 mm, respectively). However, marketable yields with full irrigation (Figure 1) were higher in Foggia than in Viterbo (95 and 59 Mg fresh weight ha^{-1}). This result can be ascribed firstly to the lower fertilization supplied at Viterbo (average N fertilization was about two thirds compared to the amount supplied at Foggia), and secondly to other environmental conditions that can positively or negatively affect crop yields (e.g., air temperature). In fact, average temperature is 15.8 °C at Foggia and climate is sub-humid; at Viterbo temperature is 14.4 °C and climate is humid. The same consideration is valid in relation to IWUE. In fact, average IWUE was higher at Matera (Figure 10), where temperature is 16.5 °C and climate is semi-arid (Figure 11).

Generally, our data-analysis has confirmed that results are quite variable and strongly site-dependent due to different climate and soil conditions that may mask the actual effect of the irrigation regime, and consequently cannot be generalized. Based on the field experiments considered, a limited decrease in water restoration according to the calculated interpolating function (Figure 5) would marginally decrease yield by 17.9% and 11.7%, at 70 and 80% of maximum water restoration, respectively. Marketable yields did not differ significantly at 40–60% and 60–80% of maximum water restoration (Figure 3) but were higher when 80–100% of maximum water restoration was supplied with DI, in agreement with previous research [20]. In addition, variable irrigation regimes during the crop cycle showed a higher and significant response to crop yields (Figure 4). Yield responses to DI were significantly lower in soils with coarse and medium textures (Figure 7). Irrigation water use efficiency was weakly significantly different among irrigation classes and water regimes but was significantly higher in the experiment conducted at Matera (Figure 10), coupled with semiarid climate conditions (Figure 11).

5. Conclusions and Recommendations

Under Mediterranean conditions, water management is a crucial factor for tomato crops, due to the limited availability of water resources during the growing season, when evapotranspiration is not balanced by the moderate amount of rainfall. Therefore, in this environment, the sustainable use of water resources is a priority. A proper application of DI can save huge amounts of water, particularly in semi-arid environments where water scarcity is an increasing concern and water costs are continuously rising.

Our results provide practical guidelines for irrigation water use in processing tomato cultivation that can be easily addressed by farmers to avoid over-irrigation and to adopt reduced irrigation rates during the less sensitive growth stages. Our recommendation is that further research should address the response of crop yield under variable irrigation regimes adopting RDI, and in relation either to coarse and medium soil textures and sub-humid climate conditions that are very frequent in Mediterranean Italy.

Alternative strategies to reduce irrigation water use can be recommended, namely sensor-based irrigation scheduling [41] or partial root-zone drying [35]. However, their implementation involves higher costs for farmers in terms of irrigation equipment and management compared to deficit or regulated deficit irrigation.

Supplementary Materials: The following are available online at http://www.mdpi.com/2077-0472/9/4/79/s1, Table S1: Main details of field experiments.

Author Contributions: R.F. and C.D.B. made substantial contributions to the manuscript. R.F. performed data curation and writing—review and editing; C.D.B. performed writing—review and editing.

Funding: This research was funded by the Diverfarming project "Crop diversification and low-input farming across Europe: from practitioners' engagement and ecosystems services to increased revenues and value chain organisation", a European Union's Horizon 2020 Programme for Research and Innovation, under grant agreement no 728003.

Conflicts of Interest: The authors declare no conflict of interest.

References

1. FAO (Food and Agriculture Organization of the United Nations). FAO Statistical Yearbook 2012. Available online: http://www.fao.org/docrep/015/i2490e/i2490e00.htm (accessed on 7 April 2019).
2. Du, Y.D.; Niu, W.Q.; Gu, X.B.; Zhang, Q.; Cui, B.J. Water- and nitrogen-saving potentials in tomato production: A meta-analysis. *Agr. Water Manag.* **2018**, *210*, 296–303. [CrossRef]
3. Molden, D. *Water for Food, Water for Life: A Comprehensive Assessment of Water Management in Agriculture*; Earthscan and Colombo, International Water Management Institute: London, UK, 2007; 688p.
4. FAO (Food and Agriculture Organization of the United Nations). AQUASTAT Main Database 2016. Available online: http://www.fao.org/nr/water/aquastat/data/query/index.html (accessed on 7 April 2019).
5. Yahyaoui, I.; Tadeo, F.; Vieira, M. Energy and water management for drip-irrigation of tomatoes in a semi-arid district. *Agr. Water Manag.* **2017**, *183*, 4–15. [CrossRef]
6. Costa, J.M.; Ortuno, M.F.; Chaves, M.M. Deficit irrigation as a strategy to save water: Physiology and potential application to horticulture. *J. Integr. Plant Biol.* **2007**, *49*, 1421–1434. [CrossRef]
7. Mushtaq, S.; Moghaddasi, M. Evaluating the potentials of deficit irrigation as an adaptive response to climate change and environmental demand. *Environ. Sci. Policy* **2011**, *14*, 1139–1150. [CrossRef]
8. Ali, M.H.; Hoque, M.R.; Hassan, A.A.; Khair, A. Effects of deficit irrigation on yield, water productivity, and economic returns of wheat. *Agr. Water Manag.* **2007**, *92*, 151–161. [CrossRef]
9. Vazifedoust, M.; Van Dam, J.C.; Feddes, R.A.; Feizi, M. Increasing water productivity of irrigated crops under limited water supply at field scale. *Agr. Water Manag.* **2008**, *95*, 89–102. [CrossRef]

10. Rodriguez-Ortega, W.; Martinez, V.; Rivero, R.; Camara-Zapata, J.; Mestre, T.; Garcia-Sanchez, F. Use of a smart irrigation system to study the effects of irrigation management on the agronomic and physiological responses of tomato plants grown under different temperatures regimes. *Agr. Water Manag.* **2017**, *183*, 158–168. [CrossRef]

11. Zegbe-Domínguez, J.A.; Behboudian, M.H.; Lang, A.; Clothier, B.E. Deficit irrigation and partial rootzone drying maintain fruit dry mass and enhance fruit quality in 'Petopride' processing tomato (*Lycopersicon esculentum*, Mill.). *Sci. Hortic.* **2003**, *98*, 505–510. [CrossRef]

12. Patanè, C.; Tringali, S.; Sortino, O. Effects of deficit irrigation on biomass, yield, water productivity and fruit quality of processing tomato under semi-arid Mediterranean climate conditions. *Sci. Hortic.* **2011**, *129*, 590–596. [CrossRef]

13. Zhang, H.M.; Xiong, Y.W.; Huang, G.H.; Xu, X.; Huang, Q.Z. Effects of water stress on processing tomatoes yield, quality and water use efficiency with plastic mulched drip irrigation in sandy soil of the Hetao Irrigation District. *Agric. Water Manag.* **2017**, *179*, 205–214. [CrossRef]

14. Chen, J.L.; Kang, S.Z.; Du, T.S.; Guo, P.; Qiu, R.Q.; Chen, R.Q.; Gu, F. Modeling relations of tomato yield and fruit quality with water deficit at different growth stages under greenhouse condition. *Agr. Water Manag.* **2014**, *146*, 131–148. [CrossRef]

15. WPTC (World Processing Tomato Council). Available online: http://www.wptc.to/ (accessed on 22 March 2019).

16. ISTAT (Istituto Nazionale di Statistica). Available online: http://agri.istat.it/ (accessed on 22 March 2019).

17. Battilani, A. Processing tomato water and nutrient integrated crop management: State of art and future horizons. *Acta Hortic.* **2003**, *613*, 63–73. [CrossRef]

18. Rinaldi, M.M.; Thebaldi, M.S.; da Rocha, M.S.; Sandri, D.; Felisberto, A.B. Postharvest quality of the tomato irrigated by different irrigation systems and water qualities. *IRRIGA* **2013**, *18*, 59–72. [CrossRef]

19. Benincasa, P.; Guiducci, M.; Tei, F. The nitrogen use efficiency: Meaning and sources of variation – case studies on three vegetable crops in Central Italy. *Hort. Technol.* **2011**, *21*, 266–273. [CrossRef]

20. Fereres, E.; Soriano, M.A. Deficit irrigation for reducing agricultural water use. *J. Exp. Bot.* **2007**, *58*, 147–159. [CrossRef] [PubMed]

21. Giuliani, M.; Nardella, E.; Gagliardi, A.; Gatta, G. Deficit irrigation and partial root-zone drying techniques in processing tomato cultivated under Mediterranean climate conditions. *Sustainability* **2017**, *9*, 2197. [CrossRef]

22. Rinaldi, M.; Garofalo, P.; Vonella, A.V. Productivity and water use efficiency in processing tomato under deficit irrigation in Southern Italy. In XIII International Symposium on Processing Tomato, Sirmione, Italy. *Acta Hort.* **2015**, *1081*, 97–104. [CrossRef]

23. Ventrella, D.; Giglio, L.; Charfeddine, M.; Lopez, R.; Castellini, M.; Sollitto, D.; Castrignanò, A.; Fornaro, F. Climate change impact on crop rotations of winter durum wheat and tomato in southern Italy: Yield analysis and soil fertility. *Ital. J. Agron.* **2012**, *7*, 15. [CrossRef]

24. Ventrella, D.; Giglio, L.; Garofalo, P.; Dalla Marta, A. Regional assessment of green and blue water consumption for tomato cultivated in Southern Italy. *J. Agr. Sci.* **2018**, *156*, 689–701. [CrossRef]

25. Marouelli, W.A.; Silva, W.L.C. Water tension thresholds for processing tomatoes under drip irrigation in Central Brazil. *Irrig. Sci.* **2007**, *25*, 411–418. [CrossRef]

26. Patanè, C.; Cosentino, S.L. Effects of soil water deficit on yield and quality of processing tomato under a Mediterranean climate. *Agr. Water Manag.* **2010**, *97*, 131–138. [CrossRef]

27. Allen, R.G.; Pereira, L.S.; Raes, D.; Smith, M. Crop evapotranspiration. In *Guidelines for Computing Crop Water Requirements*; FAO: Rome, Italy, 1998; Irrigation and Drainage Paper Volume 56, p. 300.

28. Doorenbos, J.; Pruitt, W.O. *Crop Water Requirements*; FAO Irrigation and Drainage Paper 24; FAO: Rome, Italy, 1977; p. 144.

29. Cantore, V.; Lechkar, O.; Karabulut, E.; Sellami, M.H.; Albrizio, R.; Boari, F.; Stellacci, A.M.; Todorovic, M. Combined effect of deficit irrigation and strobilurin application on yield, fruit quality and water use efficiency of "cherry" tomato (*Solanum lycopersicum* L.). *Agr. Water Manag.* **2016**, *167*, 53–61. [CrossRef]

30. Giuliani, M.M.; Gatta, G.; Nardella, E.; Tarantino, E. Water saving strategies assessment on processing tomato cultivated in Mediterranean region. *Ital. J. Agron.* **2016**, *11*, 69–76. [CrossRef]

31. Lovelli, S.; Potenza, G.; Castronuovo, D.; Perniola, M.; Candido, V. Yield, quality and water use efficiency of processing tomatoes produced under different irrigation regimes in Mediterranean environment. *Ital. J. Agron.* **2017**, *12*, 17–24. [CrossRef]

32. Leogrande, R.; Lopedota, O.; Montemurro, F.; Vitti, C.; Ventrella, D. Effects of irrigation regime and salinity on soil characteristics and yield of tomato. *Ital. J. Agron.* **2012**, *7*, 50–57. [CrossRef]

33. Candido, V.; Campanelli, G.; D'Addabbo, T.; Castronuovo, D.; Perniola, M.; Camele, I. Growth and yield promoting effect of artificial mycorrhization on field tomato at different irrigation regimes. *Sci. Hortic.* **2015**, *187*, 35–43. [CrossRef]

34. Colla, G.; Casa, R.; Lo Cascio, B.; Saccardo, F.; Temperini, O.; Leoni, C. Responses of processing tomato to water regime and fertilization in Central Italy. In Proceedings of the VI International Symposium on Processing Tomato and Workshop on Irrigation and Fertigation of Processing Tomato, Pamplona, Spain. *Acta Hort.* **1999**, *487*, 531–536. [CrossRef]

35. Casa, R.; Rouphael, Y. Effects of partial root-zone drying irrigation on yield, fruit quality, and water-use efficiency in processing tomato. *J. Hortic. Sci. Biotech.* **2014**, *89*, 389–396. [CrossRef]

36. Howell, T.A.; Steiner, J.E.; Schneider, A.D.; Evertt, S.R.; Tolk, J.A. Seasonal and maximum daily evapotranspiration of irrigated winter wheat, sorghum and corn: Southern high plains. *Trans. Asae* **1997**, *40*, 623–634. [CrossRef]

37. De Martonne, E. Une nouvelle fonction climatologique: l'indice d'aridite. *Meteorologie* **1926**, *2*, 449–458.

38. Soil Survey Staff. *Keys to Soil Taxonomy*, 12th ed.; USDA-Natural Resources Conservation Service: Washington, DC, USA, 2014; p. 360.

39. Zornoza, R.; Rosales, R.M.; Acosta, J.A.; de la Rosa, J.M.; Arcenegui, V.; Faz, Á.; Pérez-Pastor, A. Efficient irrigation management can contribute to reduce soil CO_2 emissions in agriculture. *Geoderma* **2016**, *263*, 70–77. [CrossRef]

40. Cayuela, M.L.; Aguilera, E.; Sanz-Cobena, A.; Adams, D.C.; Abalos, D.; Barton, L.; Ryals, R.; Silver, W.L.; Alfaro, M.A.; Pappa, V.A.; et al. Direct nitrous oxide emissions in Mediterranean climate cropping systems: Emission factors based on a meta-analysis of available measurement data. *Agr. Ecosyst. Environ.* **2017**, *238*, 25–35. [CrossRef]

41. Zotarelli, L.; Dukes, M.D.; Scholberg, J.M.S.; Munoz-Carpena, R.; Icerman, J. Tomato nitrogen accumulation and fertilizer use efficiency on a sandy soil, as affected by nitrogen rate and irrigation scheduling. *Agr. Water Manag.* **2009**, *96*, 1247–1258. [CrossRef]

agriculture

MDPI

Article

The Efficiencies, Environmental Impacts and Economics of Energy Consumption for Groundwater-Based Irrigation in Oklahoma

Divya Handa [1], Robert S. Frazier [1], Saleh Taghvaeian [1],* and Jason G. Warren [2]

[1]   Department of Biosystems and Agricultural Engineering, Oklahoma State University, Stillwater,
    OK 74078, USA; dhanda@ostatemail.okstate.edu (D.H.); robert.frazier@okstate.edu (R.S.F.)
[2]   Department of Plant and Soil Sciences, Oklahoma State University, Stillwater, OK 74078, USA;
    jason.warren@okstate.edu
*   Correspondence: saleh.taghvaeian@okstate.edu; Tel.: +1-405-744-8395

Received: 22 December 2018; Accepted: 27 January 2019; Published: 1 February 2019

Abstract: Irrigation pumping is a major expense of agricultural operations, especially in arid/semi-arid areas that extract large amounts of water from deep groundwater resources. Studying and improving pumping efficiencies can have direct impacts on farm net profits and on the amount of greenhouse gases (GHG) emitted from pumping plants. In this study, the overall pumping efficiency (OPE), the GHG emissions, and the costs of irrigation pumping were investigated for electric pumps extracting from the Rush Springs (RS) aquifer in central Oklahoma and the natural gas-powered pumps tapping the Ogallala (OG) aquifer in the Oklahoma Panhandle. The results showed that all electric plants and the majority of natural gas plants operated at OPE levels below achievable standard levels. The total emission from the plants in the OG region was 49% larger than that from plants in the RS region. However, the emission per unit irrigated area and unit total dynamic head of pumping was 4% smaller for the natural gas plants in the OG area. A long-term analysis conducted over the 2001–2017 period revealed that 34% and 19% reductions in energy requirements and 52% and 20% decreases in GHG emissions can be achieved if the OPE were improved to achievable standards for plants in the RS and OG regions, respectively.

Keywords: pumping plants; energy audit; life cycle assessment; greenhouse gas emission; center-pivot irrigation

1. Introduction

Irrigated agriculture around the world relies heavily on energy resources to extract freshwater and to convey it to application sites. This is especially the case in arid/semi-arid regions, where large amounts of irrigation water are required to sustain crop production. As a result, the availability and cost of energy are among major factors influencing the economic viability of irrigated agriculture in these regions. In addition, energy consumption for irrigation has major environmental consequences, mainly due to the emission of greenhouse gases [1,2]. Wang et al. [3] have reported that pumping groundwater for irrigation accounts for 3% of total emissions from agriculture in China. A similar study in Iran has found that groundwater pumping is responsible for 3.6% of total carbon emissions in the country [4]. In India, groundwater pumping is estimated to be the source of nearly 6% of India's total emissions [5]. In the U.S., carbon emissions due to pumping irrigation water have been reported to be about three million metric tons per year, with electric pumps responsible for 46% of the total emission, followed by diesel (32%) and natural gas (19%) [6].

Energy consumption and its associated costs and greenhouse gas emissions can be reduced by improving pumping efficiency [7]. In a study in central Tunisia, improving pumping efficiency was found to result in 33% cost reduction on average [8]. An average cost saving of 17% following efficiency improvement was also reported in [9] for an irrigated area in southeastern Spain. Pump efficiency is primarily dependent on operating conditions such as total dynamic head (TDH) and the condition of the pump. Any deviation from optimum conditions can lead to reduced efficiency and increased costs and emissions.

One deviation from optimum conditions is the change in TDH, caused by declines in groundwater levels. This is especially the case in irrigated areas that experience large declines due to increasing groundwater extraction. In the North China Plain, Qui et al. [10] estimated that groundwater declines from 1996 to 2013 have led to a 22% increase in energy consumption and a 42% increase in greenhouse gas emissions. Increases in groundwater depth will not only increase TDH and consequently energy use, but will also result in a gradual deviation from design parameters used in selecting the most efficient pump and hence a reduction in system efficiency.

Irrigated agriculture in Oklahoma has been facing similar energy-related challenges. In 2013, Oklahoma producers spent over USD 22 million to power more than 5,300 pumps [11]. Electricity was the main source of pumping energy, supplying water to 46% of all irrigated areas in the state. This was closely followed by natural gas, which powers pumps to irrigate 42% of all irrigated lands [11]. The remainder of irrigation pumps typically use diesel or propane units. In addition, Oklahoma producers who rely on groundwater resources have been experiencing a decline in water availability, reflected in a reduction in average well flow rates from 0.032 m^3 s^{-1} in 2008 to 0.026 m^3 s^{-1} in 2013 [11]. The groundwater decline has been significant in the Panhandle region due to increased usage, drought periods, and negligible recharge rates. Identifying energy consumption efficiencies and improved practices can have a considerable impact on the profitability of agricultural production in Oklahoma.

The goal of this study was to identify the overall efficiency and environmental impact of irrigation pumping in two agricultural regions of central and western Oklahoma that rely on two aquifers with significantly different depths to groundwater. The more specific objectives included: (i) to estimate the overall pumping efficiency for several pumping plants in each region; (ii) to conduct a life cycle assessment and calculate greenhouse gas emissions under existing and achievable efficiencies; and (iii) to investigate the impacts of changing groundwater depths and energy prices on the economics and emissions of irrigation pumping plants over a 17-year period. To the best of our knowledge, only a few previous studies have identified the overall efficiency of agricultural irrigation pumping. For example, Luc [8] has determined the efficiency of 18 electric pumps in central Tunisia. The lack of data in this field is most probably due to large human, technical, and financial resources required to carry out field evaluations of pumping plants. As a result, many previous life cycle assessment studies have assumed or approximated pumping efficiencies in their analysis as opposed to using measured values [2–4,7,10]. This study combines efficiencies measured through field audits with greenhouse gas analysis and uses the results to explore long-term effects of changing groundwater levels on energy costs and emissions. The results will give Oklahoma agricultural producers, water managers, and other decision makers an insight into current economics and the environmental footprint of irrigation pumping in the study areas, as well as potentials for improvement. Moreover, the results will be transferable to areas with similar agro-climatological and groundwater resources conditions.

2. Materials and Methods

2.1. Study Area

A total of 24 irrigation pumping plants in central and Panhandle regions of Oklahoma were tested between 2015 and 2018 with the aim of determining their energy consumption efficiencies, emissions, and expenses. Of the pumping plants evaluated, fourteen were located within the Ogallala (OG) aquifer and ten within the Rush Springs (RS) aquifer (Figure 1). The OG sites were all natural gas internal combustion powered and the RS sites were electricity-powered pumping plants. The Ogallala is one of the most important aquifers in Oklahoma, supplying more than 98% of total water demand in the Panhandle regions. It is classified as an unconfined bedrock aquifer composed of semi-consolidated clay, silt, sand, and gravel layers [12]. The maximum thickness of the aquifer is 213 m and the groundwater flow direction is toward the east/southeast, similar to the land surface elevation gradient. OG water quality is considered good, with an average pH of 7.3 and specific conductance of 0.64 decisiemens per metre (dS m^{-1}) [12]. RS is another important bedrock aquifer in the state and provides irrigation water to numerous fields in central Oklahoma. This aquifer is composed of Rush Springs sandstone on top of the Marlow formation. The maximum thickness of RS is 101 m, with groundwater moving in a south/southeast direction [12]. Similar to OG, the water quality of RS is good, with an average pH of 7.2 and specific conductance of 1.08 dS m^{-1} [12]. The depth to groundwater, however, is much larger in the OG (with a 2018 average of 57.1 m), and it has experienced a steady decline over the past several decades, while RS is shallower (with a 2018 average of 18.2 m) and is more sensitive to inter-annual variations in precipitation [11,12].

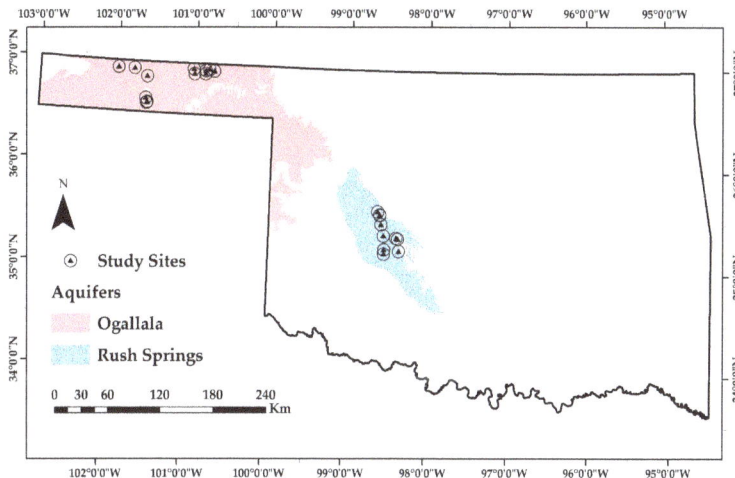

Figure 1. Location of tested pumping plants and their corresponding aquifers.

2.2. Energy Audits

The energy audits included determining several basic irrigation well and pump parameters such as groundwater depth (GWD), water pressure, and discharge rate. These parameters were then used to estimate the overall pumping efficiency (OPE), a widely used metric for assessing the efficiency of irrigation pumping plants. The OPE is the ratio of the output work the pump exerts on the water at the pump outlet, known as water horsepower (WHP), to the required energy input or energy horsepower (EHP) of the driving unit required to pump the measured water output [13], and is calculated as

$$OPE = \frac{Water\ output\ (WHP)}{Energy\ input\ (EHP)} \times 100 \tag{1}$$

The WHP (kW) can be determined as

$$WHP = \frac{Q \times TDH}{F} \tag{2}$$

where Q is the discharge rate (m^3 s^{-1}), TDH is the total dynamic head (m) and F is a conversion factor equal to 0.102 (m^4 s^{-1} kW^{-1}). In this study, Q was measured using an ultrasonic flow meter (Portaflow-C, Fuji Electric Co., Japan) on the discharge pipe from the pump. The accuracy of the ultrasonic flow meter was tested previously against a calibrated flow device and found to be acceptable [14].

The TDH is the total equivalent pressure that must be applied to the water column being pumped while also taking into account the losses due to friction [13]. It may be expressed as

$$TDH = \text{pumping lift} + \text{pressure head} \tag{3}$$

where pumping lift is the vertical distance between the pumping water level and center of the pump outlet (m) and pressure head is the pressure required at the pump outlet (m). The pumping lift was measured by lowering a water level meter (model 102, Solinst Canada Ltd., Canada) probe through an access hole in the pump base-plate while a pressure gauge close to the pump outlet was used to measure the pressure head [15].

The estimation procedure for energy horsepower (EHP) depends on the type of energy used and differs among electric motors in the Rush Springs aquifer region and natural gas engine driven pumps in the Ogallala aquifer region.

2.2.1. Electric Motors

In Oklahoma, electric motor driven irrigation pumps tend to be used where the ground water depth is less than 80 m and three-phase power is available. These pumps usually require less maintenance and operational activity than internal combustion engines. For electric motors, the energy input (kW) is the electrical power supplied to the motor and can be calculated using the following equation for a three-phase motor [16]:

$$EHP = \frac{V \times I \times PF \times 1.732}{1000} \tag{4}$$

where V is voltage (V), I is current (A), PF is the power factor, and 1.732 is a conversion factor. In this study V, I, and PF were measured using a three-phase electric meter. The current of each of the three legs was first measured individually and then averaged together. The voltage was measured across all three legs and also averaged.

2.2.2. Natural Gas Engines

The natural gas consumption of the internal combustion engines was measured by a rotary gas meter (Dresser Roots® Series B, General Electric, Boston, MA, USA). The gas meter was installed by first turning off the gas supply to the engine at the gas meter. Next, the main fuel line running to the intake manifold was disconnected and the rotary meter was installed in-line with this gas line, which was then reconnected to the engine. The engine was allowed to run until in steady state operating temperature. The irrigation water pump was also allowed to bring the entire irrigation system up to operating pressure (with water delivery from all nozzles). Then the engine and pump system were allowed to run for 30–45 min, at which time average fuel consumption readings and correction factors were recorded. Removing the rotary meter was the reverse of the installation procedure. The meter auto-corrects for gas pressure, density, and temperature. The display gives readings of cubic feet per minute, which were converted to mechanical power and used as the input fuel power.

The estimated OPE of the audited pumping plants was compared against the widely used Nebraska Pumping Plant Performance Criteria (NPPPC). According to the NPPPC, the OPE of accurately designed and appropriately maintained electricity- and natural gas-driven pumping plants should be 66% and 17%, respectively [16]. With the electric pumping plants, their OPE was also compared against thresholds proposed by Hanson [17] for different required corrective actions.

2.3. Life Cycle Assessment

The greenhouse gas (GHG) emissions for the pumping sites were calculated using the GREET® (GREET.NET version 2017, Argonne National Laboratory, Argonne, IL, USA) and the U.S. Environmental Protection Agency (EPA) Greenhouse Gas Inventory Guidance [18]. The GREET Life Cycle Assessment (LCA) tool is an online resource that is well established and supported in the LCA focus area. While the GREET model is basically a transportation LCA tool used to model vehicle emissions due to different energy sources (e.g., biofuels, natural gas, electricity, etc.), it can also provide a good approximation for stationary power systems. The GREET model examines fuels and energy production from raw materials (e.g., coal or crude oil), extraction and processing (the "well"), to storage (the "pump"). This technique in GREET is called "well to pump" (WTP).

In the case of electricity where the end-use is essentially emission-free for both stationary and vehicle uses, the WTP model can be used alone for stationary irrigation pumping plants without the need for any additional modification. The "U.S. Central and Southern Pains" Utility Mix category was chosen to represent the electrical grid generator composition for Oklahoma. In all cases, the emissions were calculated for 1000 h of irrigation system operation. In the case of natural gas, there are on-site emissions due to the engine combustion process. Therefore, the GREET WTP analysis (extraction, transmission, and distribution) was added to an established EPA stationary engine emissions calculation technique [18] to give an approximation of the total GHG emissions for stationary engines from raw material extraction to end-use combustion. In short, the electricity production generates its GHGs at different points from the natural gas engines. Both generate GHGs in preparing the fuel/energy, but the natural gas engines contribute large percentages of GHGs at the end-use (combustion) phase. Of course, current electrical production also generates combustion GHGs at the generation phase depending on the generator technology mix. GREET captures this in the electrical WTP technique.

The EPA end-use GHG estimation for the natural gas methodology employed was based on using the natural gas volume consumed (measured) during the audit field tests with a gas flow meter. The methodology is may be given as

$$Em = Fuel \times HHV \times EF \qquad (5)$$

where Em is the mass of CO_2, CH_4, or N_2O emitted, Fuel is the mass or volume of fuel combusted, HHV is the fuel heat content (higher heating value) in units of energy per mass of fuel, and EF is the emission factor of CO_2, CH_4, or N_2O per energy unit. The HHV and EF values reported in [18] for natural gas combustion were used in this study. For the total GHG emissions from combustion, CO_2 equivalence factors of 25 for CH_4 and 298 for N_2O were applied. As with electric motor pumps, the emissions for natural gas pumping engines were reported for 1000 h of irrigation system operation. Readers should be aware that the LCA approach implemented in this study has uncertainties beyond those caused by errors in the input data, mainly due to assumptions and simplifications adopted in the procedure.

2.4. Long-Term Trends

Irrigated agriculture in the study area relies heavily on groundwater. Hence, it is of great importance to investigate the impacts of long-term fluctuations in groundwater levels on the efficiencies, emissions, and economics of irrigation pumping. The first step to conducting this analysis was to estimate variations in energy requirement in response to changes in groundwater depth for each of the studied aquifers (Ogallala and Rush Springs). Several previous studies have investigated energy required for pumping groundwater as a function of depth to groundwater [4,7,19,20]. These studies have used the following equation or a variation of it; it has been selected in the present study and applied to estimate annual energy requirements over the 17-year period from 2001 to 2017:

$$Energy = \frac{TDH \times M \times g}{3.62 \times 10^6 \times OPE} \tag{6}$$

where energy is in kWh, M is the total mass of groundwater pumped for irrigation (kg), g is the gravitational acceleration (9.8 m s^{-2}), and other parameters are as they were defined above.

Since actual long-term TDH data for audited systems were not available, this parameter was approximated through developing a linear regression model to predict TDH from groundwater depth based on the data collected during energy audits. The assumption was that GWD is by far the largest portion of TDH, especially since all tested center pivot systems were of mid-elevation spray application type and thus required significantly lower operating pressures compared to center pivots equipped with impact sprinklers. The close proximity of irrigation wells to irrigation systems meant that pressure losses during water conveyance were small too. Once this relationship was developed it was applied to the average annual GWD estimated during the 2001–2017 period based on readings reported by the Oklahoma Water Resources Board at 42 and 22 observation wells in the Ogallala and Rush Springs aquifers, respectively. The readings were made on an annual basis (in winter) using electric tapes. The observation wells were selected in a way as to to represent the entire aquifer. The number of wells with continuous GWD data dropped quickly for years before 2001, which is why the 17-year period from 2001 to 2017 was selected for analysis in this study.

The discharge rates obtained during the energy audits were averaged for each studied aquifer and used in obtaining M, assuming 1000 h of pump operation per year. The OPE was estimated in a similar fashion, assuming that the average OPE of audited systems in each region was a reasonable representative of the average OPE of all systems in that region. In addition, this average OPE was assumed to be an acceptable representative of average OPE over the long-term period considered (2001–2017). After obtaining annual energy requirements, the LCA analysis explained in the previous section was implemented to estimate variations in GHG emissions as impacted by fluctuations in groundwater depth.

2.5. Economic Analysis

The economic implications of improving OPE at studied pumping plants was also investigated. Irrigation energy costs can be one of the largest categories of costs a producer in Oklahoma will incur over a season, reaching about 22 million USD [11]. Improving OPE to the levels recommended by the NPPPC could decrease irrigation pumping costs. In this study, energy cost analysis was conducted in two parts. The first part focused on estimating the energy cost of pumping 1000 h per year under current efficiencies and potential savings if the OPE of each tested pumping plant was improved to NPPPC levels. Unit energy costs of 0.05 USD per kWh for electricity-powered and 3.30 USD per 1000 cubic feet (MCF) of fuel for natural gas-powered plants were used in the analysis based on costs reported in the U.S. Energy Administration Information web portal for Oklahoma for the year when the tests were conducted [21].

The second part examined changes in long-term (2001–2017) pumping costs for the average OPE in each aquifer region by taking into account annual variations in groundwater depth and energy costs. The cost of electricity in Oklahoma varied from 0.04 USD kWh^{-1} in 2002 to 0.06 USD kWh^{-1} in 2008 and 2014 [21]. It should be noted that the actual blended rates would be higher than the projected rates used here. Hence, the actual pumping costs for these plants will be larger than those estimated in the present study. Compared to the cost of electricity, greater inter-annual variations were observed in natural gas costs in Oklahoma, ranging from 2.94 USD MCF^{-1} in 2016 to 13.03 USD MCF^{-1} in 2008 [21]. Finally, the energy costs for all electric and natural gas pumping plants in Oklahoma were approximated by assuming the estimates for audited sites are statistically representative of all corresponding plants in the state. This assumption may not be accurate, but it provides an estimate of the magnitude of statewide energy expenses for irrigation pumping.

3. Results and Discussion

3.1. Energy Audits

The measured parameters showed significant differences among audited sites in the Rush Springs and Ogallala aquifer regions. The average static groundwater depths for instance, were 24.4 and 79.7 m for RS and OG, respectively. Barefoot [22] has tested 13 natural gas irrigation pumping plants in the Oklahoma Panhandle region (OG) and reported a similar average pumping lift of 80.4 m. The average dynamic GWD, measured 15 min after starting the pump, was 30.7 and 89.1 m for the same aquifers, respectively. The measured water pressure was larger for irrigation systems in RS, resulting in a smaller difference in TDH compared to GWD. The average TDH was 67.8 and 105.9 m for the RS and OG aquifers, respectively.

The difference in TDH was accompanied by a corresponding difference in input energy. With an average value of 270 kW (362 Hp), the input power requirement in the OG was nearly five times larger than that in the RS region, which had an average of 56 kW (75 Hp). This probably explains the preference of natural gas engines over electric motors as an energy source for powering OG pumping plants since large electric motors have specific wiring and utility constraints. The water discharge rates were similar in the two study regions, with average values of 36.2 and 36.0 $l\,s^{-1}$ for the RS and OG aquifers, respectively. The average discharge reported in [22] was 47.9 $l\,s^{-1}$ in the OG aquifer region, about 33% larger than the value found in the present study.

The overall pumping efficiency of the sites in the RS aquifer region (electricity-powered) varied from 24.9% to 62.6%. Of the ten pumping plants evaluated, seven had an OPE of less than 50%, which was proposed in [17] as the threshold below which repairing or replacing the pump should be considered. All of the systems had efficiencies smaller than the recommended OPE of 66% by the NPPPC standard. The average OPE for the RS region was 43.3%. The difference between the estimated OPE and the NPPPC standard implies that nearly 23% of electrical energy is wasted on average due to poor efficiency of the pumping plant in the RS region. The average OPE in this study compares well with the average OPE of 42.6% reported in [23] and 47.0% in [24] for pumping plants in the High Plains and Trans-Pecos areas of Texas. The range of efficiencies in [24] was also similar to that in this study, with values varying from 16.8% to 70.6%. However, DeBoer et al. [25] have reported a larger average OPE of 58% in for electricity-driven pumping plants in west central Minnesota, North Dakota, and South Dakota. The plants tested in DeBoer's study were fairly new, with 74% being less than six years old. The younger age of the pumps could be the cause of higher average efficiency.

The OPE of the natural gas-powered pumping plants in the OG aquifer region ranged from 5.7% to 21.4%. Out of 14 audited pumping plants, ten had an OPE less than the NPPPC recommended standard of 17% for natural gas internal combustion engines pumping plants. The average OPE for the OG region was 13.6%, close to average OPEs of 13.2%, 11.7%, and 13.1% as reported for natural gas-powered pumping plants in Oklahoma and Texas by Barefoot [22], Fipps et al. [23], and New and Schneider [24], respectively. The range of OPEs in [24] was 2.2–21.6%, similar to the range of OPEs estimated in the present study.

Linear regression analysis conducted on data collected at each site and the two sites combined revealed no significant relationship between OPE and the two parameters of TDH and discharge rate (*p* values larger than 0.37). This suggests that the performance of audited systems was impacted by other factors such as the type, age, and condition of the pumping plants. Small sample sizes of systems tested may have also contributed to the lack of correlation. Table 1 presents the average values of key parameters for tested plants in the two study areas.

Table 1. Average values of the main characteristics of studied pumping plants in the Rush Springs (RS) and Ogallala (OG) aquifer regions.

Parameter	RS	OG
Static groundwater depth (m)	24.4	79.7
Dynamic groundwater depth (m)	30.7	89.1
Total dynamic head (m)	67.8	105.9
Discharge ($l\,s^{-1}$)	36.2	36.0
Overall pumping efficiency (%)	43.3	13.6

3.2. Life Cycle Assessment

The LCA of electric motor pumps in the RS region examined the greenhouse gas (GHG) emissions at an electric generation station for 1000 h of pump operation. The total emissions from these pumping plants ranged from 28.4 to 52.9 and averaged 40.1 metric tons of equivalent CO_2 (t CO_2e) emissions. In order to facilitate comparison with other studies, it is useful to report emissions for the unit irrigated area and unit TDH. The pumping plants in the present study were all serving a standard-size center pivot system, with an irrigated area of about 50.8 ha per system. Hence, the average emission from electrical pumps per unit area of irrigated land and unit TDH would be 11.8 kg CO_2 e ha^{-1} m^{-1}. This is within the range of 4–93 kg CO_2 e ha^{-1} m^{-1} reported in [7] for a variety of crops irrigated by electrical pumps in the Haryana state of India.

LCA of natural gas driven pumps examined the emissions through a two-part analysis that used GREET WTP and EPA calculations for stationary engines. The first part of the analysis provided estimates for natural gas extraction, processing, storage, and transportation (off-site), while the second part resulted in emission estimates for end-use at the irrigation field (on-site). The total off-site GHG emissions estimated for natural gas pumping plants in the OG region averaged 11.0 and had a range of 6.0–17.6 t CO_2e. The on-site emissions varied from 26.7 to 78.1 and averaged 48.8 t CO_2e. The average total emission from off- and on-site analyses was 59.8 t CO_2e. This is equal to 11.4 kg CO_2e ha^{-1} m^{-1} when expressed in terms of emission per unit area of irrigated land and unit TDH. The maximum total emission was 95.7 t CO_2e and belonged to a pumping plant that had the fourth lowest OPE and the seventh largest groundwater depths, a combination resulting in the maximum amount of energy use.

On average, the total GHG emission from pumping plants in the OG region was 49% larger than that of the pumping plants in the RS region. However, the emission per unit irrigated land and unit TDH was 4% smaller. This indicates that TDH, which is greatly impacted by groundwater depth, plays a significant role in determining the GHG emissions from agricultural pumping plants. The emissions can also be reported per unit volume of water extracted from aquifers. This analysis resulted in emission ranges of 0.20–0.45 and 0.34–0.99 kg CO_2 e m^{-3} for pumping plants in the RS and OG regions, respectively. These estimates are similar to the range of 0.18–0.60 kg CO_2 e m^{-3} reported in [3] for all 31 provinces in China with variable proportions of electricity and diesel driven pumps.

3.3. Long-Term Trends

Examination of the observed groundwater depth data showed that the Rush Springs aquifer levels varied between 18.2 and 21.0 m below the ground surface over the 17-year period, with a net decline of 1.5 m (Figure 2a). On the other hand, the Ogallala aquifer GWD experienced a steady decline from 56.6 to 62.3 m below the ground surface (Figure 2b). The OG aquifer has significantly smaller recharge rates. As a result, no rise in water level was observed in the OG aquifer during wet periods in 2005, 2007–2009, and 2015–2017, while the RS aquifer experienced rises in groundwater level. The rate of decline in water levels was greater during the drought years of 2011–2014 compared to wet and normal years for both aquifers, an indication of increased pumping for irrigation during this dry period.

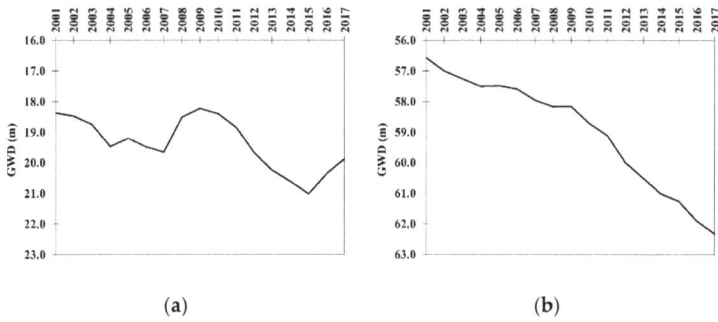

Figure 2. Annual variations in groundwater depth (GWD) for plants in the (a) Rush Springs, and (b) Ogallala aquifers.

The TDH and GWD measurements at audited sites were strongly correlated, with a Pearson coefficient of 0.88. The linear relationship developed based on TDH and GWD is presented below:

$$TDH = 0.67 \times GWD + 53.76, \tag{7}$$

This relationship was statistically significant ($p < 0.001$) and had a coefficient of determination (R^2) of 0.78, suggesting that over three-fourths of variability in TDH could be explained by changes in GWD. A similar approach was employed in [3], where the slope, intercept, and R^2 were 0.91, 21.75, and 0.62 for a linear relationship between pump lift and GWD.

As expected, the variations in energy requirement during the 2001–2017 period had a pattern similar to that of GWD in each aquifer region. In the case of RS, the energy requirement for 1000 h of pump operation per year varied from 53,721 to 55,247 kWh during the 17 years considered and had an average of 54,344 kWh. The energy requirement was much larger at OG and increased over time, with a range of 233,175–242,980 kWh and an average of 237,277 kWh (Figure 3). This was more than four times larger than the average in RS. When considering energy requirements per unit volume of pumped water, the RS and OG regions had average rates of 0.42 and 1.84 kWh m^{-3}, respectively. These values are similar to energy use rates of 0.21 to 0.66 kWh m^{-3} reported in [3] across all provinces in China.

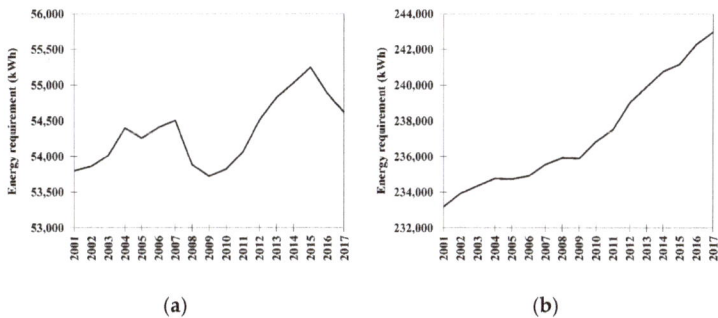

Figure 3. Annual variations in energy requirements for 1000 h of operation per year for plants in the (a) Rush Springs, and (b) Ogallala aquifers.

In OG, the increase in energy requirements due to increases in GWD over the 17-year period was 4% of the initial (2001) amount. Qiu et al. [10] have reported a significantly larger increase of 22% in energy use in China between 1996 and 2013. However, the rate of groundwater level decline in their study was 0.6 m year^{-1}, two times larger than the drop rate of 0.3 m year^{-1} observed in the present study. The results also revealed that improving the OPE in each region to achievable levels recommended by the NPPPC would result in 34% and 19% reductions in average energy requirements in the RS and OG regions, respectively.

The increase in energy requirements in the OG aquifer region over the long-term period resulted in a continuous increase in total GHG emissions from 51.7 t CO_2e in 2001 to 53.9 t CO_2e in 2017. In the RS aquifer region, the total GHG emissions varied between 38.3 t CO_2e in 2009 and 39.4 t CO_2e in 2015, the year that marked the end of a severe drought that caused significant declines in groundwater levels. Apart from the groundwater level, the OPE had a large influence on the energy use rate and emissions. Improving the OPE of electricity-powered pumping plants in the RS aquifer region to the NPPPC recommended standard of 66% could on an average reduce total GHG emissions by nearly 52%. Similarly, improving the OPE of natural gas-powered pumping sites from an average of 13.75% to the NPPPC recommended 17% level in the OG aquifer region could potentially reduce emissions by 20%. In India, it has been reported that improving electric pumping system efficiency (OPE) to 51% from an existing 34.7% level could lead to a decline of 32% in CO_2 emissions [7].

3.4. Economic Analysis

Based on the audit results, the current seasonal pumping cost for 1000 h was highly variable among pumping plants, with an average of 2827 USD for electricity-powered pumping sites in the RS aquifer region. The pumping expense was also variable among natural gas-powered pumping sites, with an average of 3042 USD, which was only 8% larger than that of electric plants. Such a small difference despite significant differences in energy use is due to low natural gas prices during the study period. It should be also noted that producers may run their pumps longer than 1000 h per year depending on crop type, weather conditions, and well yield. A longer operating hour will result in a linear increase in energy consumption costs. A significant potential for reduction in pumping costs was estimated if OPE was improved to meet NPPPC standards. The average saving for electric pumps was 35% of the current pumping costs. With the prices at the time of study, this amount of saving was equal to 1000 USD or one USD per every hour of pumping. For natural gas-powered pumping plants, the average saving was 23% of existing pumping costs, equal to 711 USD (71 cents per every hour of pumping). As mentioned before, this study was conducted when natural gas prices were among the smallest in the past several years in Oklahoma. Higher costs would have resulted in larger dollar values of saving. The results obtained here are consistent with the findings of Hardin and Lacewell [26] who reported significant decreases in fuel costs and increases in farm profits if the OPE of irrigation pumping plants in the Texas High Plains was improved to achievable levels.

Our long-term analysis showed that pumping cost for electric plants ranged from 2052 USD in 2002 to 3219 USD in 2014 and averaged 2788 USD for 1000 h of pumping per year (Figure 4). In the case of natural gas, inter-annual fluctuations were significantly larger. The pumping costs for these plants varied between 2429 USD in 2016 and 10,482 USD in 2008 and averaged 6490 USD for 1000 h of pumping per year (Figure 4). According to the most recent Farm and Ranch Irrigation Survey (FRIS) conducted by the U.S. Department of Agriculture, 3456 electricity-powered and 1354 natural gas-powered pumping plants are used for irrigation in Oklahoma [27]. Assuming that these plants have pumping costs similar to those estimated in this study, the total annual cost of irrigation pumping in Oklahoma can be approximated at 9.6 and 8.8 million USD for electric and natural gas plants, respectively. These estimates are similar to the total energy expenditure of 9.5 and 10.5 million USD reported for the same two sources of energy in [27], respectively.

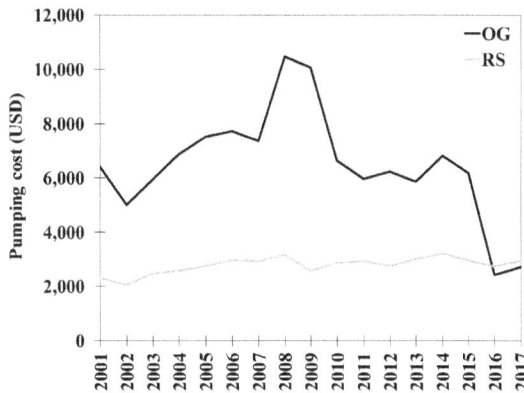

Figure 4. Annual variations in pumping cost for 1000 h of operation per year for the Ogallala and Rush Springs aquifers.

4. Conclusions

The future of irrigated agriculture is largely dependent on its financial and environmental sustainability. In arid/semi-arid regions where large amounts of irrigation water is required to meet crop demand and where available water resources are usually limited or difficult to extract, energy consumption for irrigation pumping can play a significant role in both environmental and financial sustainability of agricultural production. In this study, the efficiencies, environmental footprint, and economics of irrigation pumping were investigated in two Oklahoma areas that rely heavily on groundwater resources. The first area was in central Oklahoma where electric pumps are used to extract water from the Rush Springs aquifer and the second area was in the Oklahoma Panhandle where natural gas-powered pumps remove water from the declining Ogallala aquifer. Field visits were conducted in the period from 2015 to 2018 to collect required data for analysis. The results confirmed significant differences among the regions in terms of depth to groundwater. It was also revealed that all electric pumps and the majority of natural gas plants had an overall pumping efficiency below the standard rates achievable under field conditions. The range and average OPE obtained in this study were similar to those reported in previous studies, including those that were conducted a few decades ago.

The life cycle assessment results showed a significant difference in total emissions, with natural gas-powered plants emitting 49% more greenhouse gases when considering the entire process from the extraction and transportation of natural gas to its combustion at the pump site. However, the average emission expressed per unit irrigated area and unit total dynamic head of pumping was slightly smaller for the natural gas plants in the OG region. In the same region, the energy required for pumping increased by 4% between 2001 and 2017 due to continuous declines in OG water levels. The results showed that improving the OPE to achievable standards could have resulted in 34% and 19% reductions in average energy consumption during the 2001–2017 period in the RS and OG regions, respectively. The reductions in GHG emissions would have been 52% and 20% for the same study areas, respectively.

The cost of irrigation pumping was also estimated for the pumping plants tested and for the two regions over the 2001–2017 period. Compared to RS, large inter-annual variations in the cost of pumping were found for agricultural producers in the OG, mainly due to the large fluctuations in the natural gas price. Assuming that the pumping plants audited in the present study represent all electric and natural gas plants in Oklahoma, the statewide energy expenses were estimated and found to be in good agreement with those reported in surveys conducted by the U.S. Department of Agriculture. The results of this study highlight the need for regular evaluation of pumping plant efficiencies to identify systems with poor performance and taking corrective actions to improve farm profitability and reduce environmental consequences of energy consumption for irrigation extraction.

Author Contributions: Conceptualization, R.S.F. and S.T.; Methodology, R.S.F. and S.T.; Formal analysis, D.H., R.S.F., and S.T.; Investigation, D.H., R.S.F., S.T., and J.G.W.; Writing—original draft preparation, D.H., R.S.F., S.T., and J.G.W.; Writing—review and editing, D.H., R.S.F., S.T., and J.G.W.; Project administration, R.S.F.; Funding acquisition, R.S.F.

Funding: This research was funded by a joint research and extension program funded by the Oklahoma Agricultural Experiment Station (Hatch funds) and Oklahoma Cooperative Extension (Smith-Lever funds) received from the National Institute for Food and Agriculture, U.S. Department of Agriculture. Additional support was provided by the Oklahoma Water Resources Center through the U.S. Geological Survey 104(b) grants program.

Acknowledgments: The authors are grateful to Mr. Chris Stoner and Mr. Donald Sternitzke from the Oklahoma Natural Resources Conservation Service, U.S. Department of Agriculture, for their support. We are also thankful to all agricultural producers who collaborated with us on this project.

Conflicts of Interest: The authors declare no conflict of interest.

References

1. Khan, M.A.; Khan, M.Z.; Zaman, K.; Naz, L. Global estimates of energy consumption and greenhouse gas emissions. *Renew. Sustain. Energy Rev.* **2014**, *29*, 336–344. [CrossRef]
2. Pradeleix, L.; Roux, P.; Bouarfa, S.; Jaouani, B.; Lili-Chabaane, Z.; Bellon-Maurel, V. Environmental impacts of contrasted groundwater pumping systems assessed by life cycle assessment methodology: Contribution to the water-energy nexus study. *Irrig. Drain.* **2015**, *64*, 124–138. [CrossRef]
3. Wang, J.; Rothausen, S.G.; Conway, D.; Zhang, L.; Xiong, W.; Holman, I.P.; Li, Y. China's water–energy nexus: Greenhouse-gas emissions from groundwater use for agriculture. *Environ. Res. Lett.* **2012**, *7*, 014035. [CrossRef]
4. Karimi, P.; Qureshi, A.S.; Bahramloo, R.; Molden, D. Reducing carbon emissions through improved irrigation and groundwater management: A case study from Iran. *Agric. Water Manag.* **2012**, *108*, 52–60. [CrossRef]
5. Shah, T. Climate change and groundwater: India's opportunities for mitigation and adaptation. *Environ. Res. Lett.* **2009**, *4*, 035005. [CrossRef]
6. Follett, R.F. Soil management concepts and carbon sequestration in cropland soils. *Soil Tillage Res.* **2001**, *61*, 77–92. [CrossRef]
7. Patle, G.T.; Singh, D.K.; Sarangi, A.; Khanna, M. Managing CO_2 emission from groundwater pumping for irrigating major crops in trans indo-gangetic plains of India. *Clim. Chang.* **2016**, *136*, 265–279. [CrossRef]
8. Luc, J.P.; Tarhouni, J.; Calvez, R.; Messaoud, L.; Sablayrolles, C. Performance indicators of irrigation pumping stations: Application to drill holes of minor irrigated areas in the Kairouan plains (Tunisia) and impact of malfunction on the price of water. *Irrig. Drain.* **2006**, *55*, 85–98. [CrossRef]
9. Mora, M.; Vera, J.; Rocamora, C.; Abadia, R. Energy efficiency and maintenance costs of pumping systems for groundwater extraction. *Water Resour. Manag.* **2013**, *27*, 4395–4408. [CrossRef]
10. Qiu, G.Y.; Zhang, X.; Yu, X.; Zou, Z. The increasing effects in energy and GHG emission caused by groundwater level declines in North China's main food production plain. *Agric. Water Manag.* **2018**, *203*, 138–150. [CrossRef]
11. Taghvaeian, S. Irrigated Agriculture in Oklahoma. Oklahoma Cooperative Extension, Publication BAE-1530. 2014. Available online: http://pods.dasnr.okstate.edu/docushare/dsweb/Get/Document-9561/BAE-1530web.pdf (accessed on 10 October 2018).
12. Oklahoma Water Resources Board. Oklahoma Groundwater Report. Beneficial Use Monitoring Program. 2017. Available online: https://www.owrb.ok.gov/quality/monitoring/bump/pdf_bump/Reports/GMAPReport.pdf (accessed on 10 October 2018).
13. Brar, D.; Kranz, W.L.; Lo, T.H.; Irmak, S.; Martin, D.L. Energy conservation using variable-frequency drives for centerpivot irrigation: Standard systems. *Trans. ASABE* **2017**, *60*, 95–106.
14. Masasi, B.; Frazier, R.S.; Taghvaeian, S. Review and Operational Guidelines for Portable Ultrasonic Flowmeters. Oklahoma Cooperative Extension, Publication BAE-1535. 2017. Available online: http://pods.dasnr.okstate.edu/docushare/dsweb/Get/Document-10723/BAE-1535web.pdf (accessed on 10 October 2018).

15. Frazier, R.S.; Taghvaeian, S.; Handa, D. Measuring Depth to Groundwater in Irrigation Wells. Oklahoma Cooperative Extension, Publication BAE-1538. 2017. Available online: http://pods.dasnr.okstate.edu/docushare/dsweb/Get/Document-10865/BAE-1538web.pdf (accessed on 10 October 2018).

16. Ross, E.A.; Hardy, L.A. *National Engineering Handbook; Irrigation Guide*; USDA: Beltsville, MD, USA, 1997.

17. Hanson, B. Improving pumping plant efficiency does not always save energy. *Calif. Agric.* **2002**, *56*, 123–127. [CrossRef]

18. U.S. Environmental Protection Agency. *Greenhouse Gas Inventory Guidance*; United States Environmental Protection Agency: Washington, DC, USA, 2016.

19. Rothausen, S.G.; Conway, D. Greenhouse-gas emissions from energy use in the water sector. *Nat. Clim. Chang.* **2011**, *1*, 210–219. [CrossRef]

20. Shahdany, S.M.H.; Firoozfar, A.; Maestre, J.M.; Mallakpour, I.; Taghvaeian, S.; Karimi, P. Operational performance improvements in irrigation canals to overcome groundwater overexploitation. *Agric. Water Manag.* **2018**, *204*, 234–246. [CrossRef]

21. U.S. Energy Information Administration. Available online: https://www.eia.gov/ (accessed on 10 October 2018).

22. Barefoot, A.D. *Investigation of Factors Affecting Energy for Irrigation Pumping*; Oklahoma Water Resources Research Institute: Stillwater, OK, USA, 1980.

23. Fipps, G.; Neal, B. *Texas Irrigation Pumping Plant Efficiency Testing Program*; Texas A & M University System: College Station, TX, USA, 1995.

24. New, L.; Schneider, A.D. *Irrigation Pumping Plant Efficiencies, High Plains and Trans-Pecos Areas of Texas*; Texas Agricultural Experiment Station, Texas A & M University System: College Station, TX, USA, 1988.

25. DeBoer, D.W.; Lundstrom, D.R.; Wright, J.A. Efficiency analysis of electric irrigation pumping plants in the upper Midwest, USA. *Energy Agric.* **1983**, *2*, 51–59. [CrossRef]

26. Hardin, D.C.; Lacewell, R.D. Implication of improved irrigation pumping efficiency for farmer profit and energy use. *J. Agric. Appl. Econ.* **1979**, *11*, 89–94. [CrossRef]

27. U.S. Department of Agriculture 2013 Farm and Ranch Irrigation Survey. Available online: https://www.nass.usda.gov/Publications/AgCensus/2012/Online_Resources/Farm_and_Ranch_Irrigation_Survey/ (accessed on 10 October 2018).

agriculture

MDPI

Article

Effect of Irrigation Water Regimes on Yield of *Tetragonia Tetragonioides*

Gulom Bekmirzaev [1,*], Jose Beltrao [2] and Baghdad Ouddane [3]

[1] Tashkent Institute of Irrigation and Agricultural Mechanization Engineers, Department of Irrigation and Melioration, Kori-Niyoziy street 39, 100000 Tashkent, Uzbekistan

[2] Research Centre for Spatial and Organizational Dynamics, University of Algarve, Campus de Gambelas, 8005-139 Faro, Portugal; jbeltrao@ualg.pt

[3] Physico-Chemistry Team of the Environment, Sciences and Technologies, University of Lille, LASIR UMR-CNRS 8516, Building C8, 59655 Villeneuve d'Ascq, CEDEX, France; baghdad.ouddane@univ-lille1.fr

* Correspondence: gulombek@gmail.com

Received: 26 September 2018; Accepted: 24 December 2018; Published: 15 January 2019

Abstract: The main purpose of this experiment was to study the effect of several irrigation water regimes on *Tetragonia tetragonioides* (Pall) O. Kuntze in semi-arid regions. During the experiment period, it was measured that several irrigation regimes were affected in terms of growth, biomass production, total yield, mineral composition, and photosynthetic pigments. The experiment was conducted in the greenhouse at the University of Algarve (Portugal). The study lasted from February to April in 2010. Three irrigation treatments were based on replenishing the 0.25-m-deep pots to field capacity when the soil water level was dropped to 70% (T1, wet treatment), 50% (T2, medium treatment), and 30% (T3, dry treatment) of the available water capacity. The obtained results showed that the leaf mineral compositions of chloride and sodium, the main responsible ions for soil salinization and alkalization in arid and semi-arid regions, enhanced with the decrease in soil water content. However, the minimum amounts of chlorophyll, carotenoids, and soluble carbohydrates in the leaf content were obtained in the medium and driest treatments. On the other hand, growth differences among the several irrigation regimes were very low, and the crop yield increased in the dry treatment compared to the medium treatment; thus, the high capacity of salt-removing species suggested an advantage of its cultivation under dry conditions.

Keywords: irrigation water regimes; leaf mineral composition; semi-arid regions; available water capacity; biomass production; total yield

---

1. Introduction

In arid and semi-arid regions, such as the Mediterranean, supplies of good-quality water allocated to agriculture are expected to decrease because most available fresh/potable water resources were already mobilized [1]. According to the Food and Agriculture Organization (FAO) [2], due to the shortage of water, there is an enlargement of saline land in agricultural areas in some developing countries. As a result, yield is decreasing, provoking an increasing cost of agricultural products [3].

Soil salinization is recognized worldwide as being among the most important problems for crop production in arid and semi-arid regions [4]. Water deficit and salinity are the major limiting factors for plant productivity, affecting more than 10% of arable land on our planet, resulting in a yield reduction of more than 50% for most major crop plants [5]. The usually noted abiotic stresses that include a component of cellular water deficit are salinity and low temperature; stresses can also severely limit crop production [6]. Abiotic stresses, such as drought, salinity, extreme temperatures, chemical toxicity, and oxidative stress are serious threats to agriculture, and result in the deterioration of the

Agriculture **2019**, *9*, 22; doi:10.3390/agriculture9010022

www.mdpi.com/journal/agriculture

environment. Abiotic stress is the primary cause of crop loss [7,8]. This problem is intensified in coastal areas due to sea-water intrusion. This results from reduced ground-water levels as the water demand exceeds the annual groundwater recharge [9]. As reported above, some of the emerging regions in risk of increasing levels of salinization of their soils are located in the Mediterranean Basin [10,11], Australia [12], Central Asia [13], and Northern Africa [14]. Salinity is one of the rising problems causing tremendous yield losses in many regions of the world, especially in arid and semi-arid regions. The use of halophytes can be an effective way of accumulating the salt in soil [15].

Intensive irrigation of agricultural crops with a high level of water mineralization causes salts to accumulate in the root zones, which adversely affects the crop productivity. In order to reduce such negative impacts, a regulated deficit irrigation (RDI) technique was adopted to combat salinization in arid and semi-arid environments by reducing the water application during certain growth stages of the crops [16].

When RDI is not feasible, halophyte crops might be a solution for the salinization of agricultural land. These crops can be irrigated by, for example, seawater, salt-contaminated phreatic sheets, brackish water, wastewater, or drainage water from other plantations [17,18].

Hence, our aims were to choose a salt-removing crop, tolerant to salinity, along with interest as a food crop, and to test its drought tolerance through its response to several water regimes. *Tetragonia tetragonioides* was the selected crop. In a previous experiment, its capability as a high biomass horticultural leaf crop was demonstrated, producing a plant dry weight of 40,000–50,000 dry mass (DM) kg·ha^{-1} if the plant population density is around 75,000 plants·ha^{-1} [19].

2. Materials and Methods

2.1. Experimental Procedure

The experimental work was conducted in the greenhouse of Horto at the University of Algarve, Faro, Portugal (37°2'37.1 N 7°58'30.8 W), from February to April in 2010. The salt-removing species *T. tetrogonioides* was selected. Plants were transplanted to 7-L pots when they had four leaves (10 February). The number of plants per pot was three, with four replications. The species were irrigated with tap water every three days until the beginning of the treatments (1 February–8 March). A nitrogen fertigation treatment was started on 8 March, with daily applied concentrations of 2 mM NO_3^- and 2 mM NH_4^+ as the cumulative amount of NO_3NH_4 (g·plant^{-1}) to the end of the experimental studies (22 April). The electrical conductivity (EC$_w$) of irrigation water was 0.6 dS·m^{-1} and pH 7.

The treatments consisted of three irrigation regimes in a randomized complete block design with three replicated treatments based on replenishing the 0.25-m-deep pots to field capacity when the soil water level dropped to 70% (T1, wet treatment), 50% (T2, medium treatment), and 30% (T3, dry treatment) of the available water capacity (aw). This concept was developed by Reference [20], where "aw" is the range of available water that can be stored in soil and is available for growing crops. It was assumed by the same authors that the soil water content readily available to plants (θ_{aw}) is the difference between the volume of water content at field capacity (θ_{fc}) and at the permanent wilting point (θ_{wp}), calculated as follows:

$$\theta_{aw} = \theta_{fc} - \theta_{wp} \tag{1}$$

The watering volume was estimated to replenish the soil profile to field capacity at a depth of 0.25 m. The volumetric soil water content (m^3 water/m^3 soil; m^3·m^{-3}) was determined just before the water application.

To control soil water along the soil profile, the irrigation frequency, and the water amount, the pots were weighed every day. The soil water content was monitored periodically, gravimetrically measured for a depth of 0.00–0.25 m.

The plants were harvested destructively (26 April), washed in water, and dried with paper towels. Then, the fresh weight (FW) was measured. The fresh samples were dried in a forced drought oven at 70 °C for 48 h, and the dry weight (DW) was measured. Plant materials were collected for chemical analyses. The electrical conductivity (EC$_s$) and pH of soil were measured before and after the experiment.

2.2. Growth and Chemical Analysis

During the vegetation period, the stem length was measured, as well as the number of nodes and number of leaves of *T. tetragonioides* every seven days.

The plants' leaves were analyzed on total growth and mineral compositions (Na, Cl, N, K, P, Ca, and Mg). Dried leaves and stems were finally grounded and analyzed using the dry-ash method. The levels of Na and K were determined using a flame photometer, and the remaining cations (Na, K, Ca and Mg) were assessed by atomic absorption spectrometry. Chloride ions were determined in the aqueous extract by titration with silver nitrate according to the method of Reference [21]. Plant nitrogen (N) content was determined using the Kjeldhal method. Phosphorus was determined using the colorimetric method according to the vanadate–molybdate method. All mineral analyses were only performed on the leaves.

The analysis of pigments was done with a disc size of 0.66 cm and a total area of 1.37 cm^2. For sugars. there were ten discs, with a disc size of 0.66 cm and a total area of 3.42 cm^2. The amount of photosynthetic pigments (chlorophyll a (Chla), b (Chlb), total (ChlT), and carotenoids) was determined according to the method of Reference [22]. Shoot samples (0.25 g) were homogenized in acetone (80%). The extract was centrifuged at 3000× g, and absorbance was recorded at wavelengths of 646.8 and 663.2 nm for the chlorophyll assay and at 470 nm for the carotenoid assay using a Varian Cary 50 ultraviolet–visible light (UV–Vis) spectrophotometer. The levels of Chla, Chlb, ChlT, and carotenoids were calculated. Soluble sugars (glucose) in leaves were extracted as described by Reference [23]. The change in absorbance was continuously followed at 340 nm using an Anthos hat II microtiter-plate reader (AnthosLabtec Instrument, Hanau, Germany).

2.3. Statistical Analyses

Data (n = 4) were examined by one-way ANOVA. Multiple comparisons of the means of data between different salinity treatments within the plants were performed using Duncan's test at the $p < 0.05$ significance level (all tests were performed with the SPSS program version 17.0 for Windows).

2.4. Soil

Table 1 shows the soil texture and soil parameters before the experiment. According to the FAO, based on the United States Department of Agriculture (USDA) particle-size classification, the soil texture was sandy clay loam. The soil parameters show that the range in the soil's pH value was slightly alkaline and that the electrical conductivity (EC$_s$) was 1.1 dS·m^{-1} (non-saline soil) at 25 °C.

Table 1. Soil parameters before the experiment.

Soil Texture		Soil Parameters			
Sand (%)	58.9	Field capacity θ_{fc} (m^3·m^{-3})	0.22	pH (H$_2$O)	7.7
Silt (%)	18	Wilting point θ_{wp} (m^3·m^{-3})	0.12	EC$_e$* (dS·m^{-1})	1.1
Clay (%)	24.1	Available soil water θ_{aw} (m^3·m^{-3})	0.12		
Classification: Sandy clay loam		Bulk density (g·cm^{-3})	1.41		

EC$_e$*—Electrical conductivity of the extract of a saturated soil paste (dS·m^{-1}).

Table 2 shows the volumetric soil water content (m^3 water/m^3 soil; m^3·m^{-3}) just before the water application. The volumetric soil water content in soil ranged between 0.20 and 0.15 m^3·m^{-3}.

Table 2. Volumetric soil water content (m^3 water/m^3 soil; m$^3 \cdot$m^{-3}) just before the water application.

Treatment	Determination	Θ (m$^3 \cdot$m^{-3})
T1	$\theta_{wp} + 0.70 \times \theta_{aw}$	$\Theta_1 = 0.20$
T2	$\theta_{wp} + 0.50 \times \theta_{aw}$	$\Theta_2 = 0.17$
T3	$\theta_{wp} + 0.30 \times \theta_{aw}$	$\Theta_3 = 0.15$

2.5. Climate Condition in Greenhouse

The average climatic data during the experimental period in the greenhouse were as follows: maximal relative humidity, 88.4%; minimal relative humidity, 11.3%; maximal temperature, 45.8 °C; minimal temperature, 11.4 °C.

During the experimental period, the relative humidity of the greenhouse was increased, and the maximal temperature decreased.

3. Results and Discussion

3.1. Effect of Irrigation Water Regimes on Plant Growth

Table 3 shows the irrigation water regimes' effects on the *T. tetragonioides* growth (stem length, number of nodes, and number of leaves). A significant effect on the stem length can be seen. In the beginning of the experiment, the stem length of the crop showed very low variations between T1 and T2 treatments. During the last three weeks of the experimental period, the stem length increased showing equal differences between each treatment—T1 and T2, and T2 and T3 (Δ stem length ~0.5 cm). The number of nodes and number of leaves were also higher in treatment T1.

Table 3. Effect of irrigation water regimes on stem length, number nodes, and number of leaves of the species. Different letters within a column represent significant differences ($p \leq 0.05$).

Treatment	*Tetragonia tetragonioides*		
	Stem Length (cm)	Number of Nodes	Number of Leaves
T1	38.8 ± 1.9 a	22.5 ± 0.6 a	9.9 ± 0.58 a
T2	34.2 ± 0.5 b	18.1 ± 0.6 b	8.1 ± 0.37 b
T3	29.3 ± 1.5 c	19.2 ± 0.8 b	9.5 ± 0.22 b

3.2. Fresh (FW) and Dry (DW) Weight of Crop

The fresh weight (FW) of *T. tetragonioides* species showed low variation among treatments (Figure 1). There was a low increase of the fresh weight of stem, leaves, and seeds in treatment T1. Surprisingly, the obtained results in treatment T3 were slightly higher than in treatment T2.

The obtained results of dry matter show that the stem, leaves, and seeds of treatment T1 were slightly higher than other treatments. There was very low variation of dry matter between T2 and T3 treatments (Figure 2).

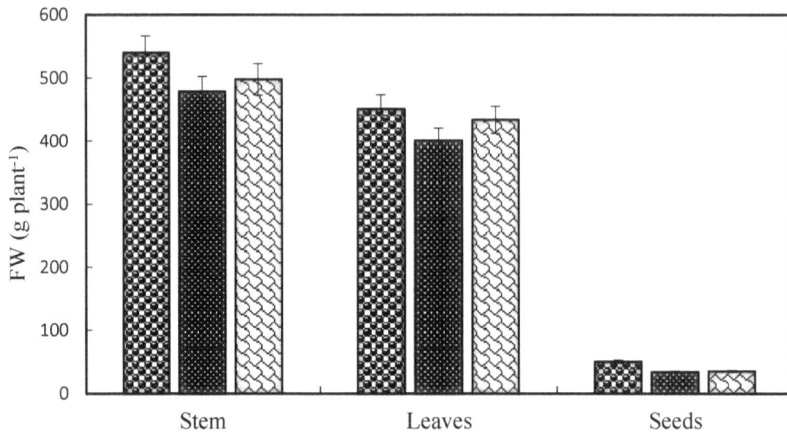

Figure 1. Fresh weight response of *Tetragonia tetragonioides* to the different irrigation treatments.

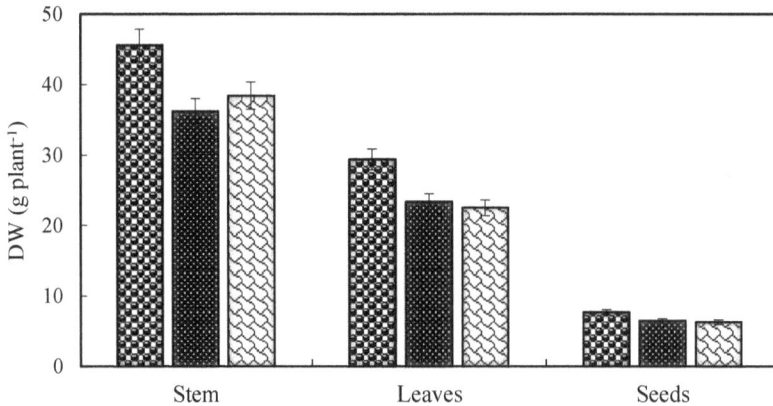

Figure 2. Dry weight response of *Tetragonia tetragonioides* to the different irrigation treatments.

3.3. Effect of Irrigation Water Regimes on Mineral Composition in Plant Leaves

Table 4 shows the effects of water application treatments on the mineral composition of *T. tetragonioides* leaves. In summary, the total nitrogen leaf content of the species showed low variation among treatments. There was an enhancement of chloride and sodium concentration with the decrease in water content. There was a general decrease in phosphorus, calcium, potassium, iron, and magnesium in the leaf content under drought conditions.

Table 4. Effect of irrigation water regimes on the leaf mineral composition.

Treatment	Leaf Mineral Composition (%)			
	Na	Cl	Mg	Ca
T1	3.4 ± 0.19	1.3 ± 0.07	0.43 ± 0.03	0.0006 ± 0.04
T2	4.3 ± 0.25	3.0 ± 0.22	0.36 ± 0.02	0.0004 ± 0.02
T3	4.4 ± 0.42	3.5 ± 0.09	0.35 ± 0.01	0.0004 ± 0.03
	N	K	P	Fe
T1	0.34 ± 0.02	4.2 ± 0.23	3.1 ± 0.02	0.0001 ± 0.02
T2	0.34 ± 0.01	3.7 ± 0.17	2.7 ± 0.04	0.0002 ± 0.03
T3	0.37 ± 0.01	4.1 ± 0.14	2.7 ± 0.02	0.0001 ± 0.01

3.4. Effect of Irrigation Water Regimes on Chlorophyll Content

The reaction of chlorophyll content of leaves of *T. tetragonioides* to the different water regimes is shown in Table 5. The results show that the chlorophyll content was higher in treatment T2 and lower in treatments T1 and T3. These results are in agreement with the findings obtained by Reference [24], where the minimum amounts of chlorophyll a, chlorophyll b, and total chlorophyll were obtained from the wettest and driest treatment in *Matricariachamomilla* L. potted plants. Similar results were obtained by References [25,26].

Table 5. Leaf chlorophyll content in leaf. DM—dry matter.

Treatment	Chlorophyll Content					
	C_a $(mg \cdot m^{-2})$	C_b $(mg \cdot m^{-2})$	C_{a+b} $(mg \cdot m^{-2})$	C_a $(mg \cdot g^{-1}; DM)$	C_b $(mg \cdot g^{-1}; DM)$	C_{a+b} $(mg \cdot g^{-1}; DM)$
T1	232 ± 12.1	81 ± 5.8	313 ± 17.6	27 ± 1.4	9.2 ± 0.7	36 ± 1.9
T2	289 ± 7.3	110 ± 4.5	400 ± 11.4	32 ± 1.3	12.2 ± 0.5	45 ± 1.7
T3	254 ± 13	92 ± 5.6	346 ± 18	30 ± 2	10.7 ± 0.6	41 ± 2.1

3.5. Effect of Irrigation Water Regimes on Carotenoid Content in Leaves

Carotenoids which exist in all higher plants are synthesized and located in the chloroplast along with the chlorophyll. Table 6 shows the carotenoid content of the leaves of *T. tetragonioides* under different irrigation water regimes. The maximum leaf carotenoid content was 8.44 mg·g^{-1} DW in treatment T2. In wetter and drier treatments (T1 and T3), the carotenoid content was lower, with values of 7.2 and 7.9 mg·g^{-1} DW, respectively. Lower carotenoid content was also obtained for stress water regimes of some fenugreek varieties [27]. Moreover, the leaf carotenoid levels of green beans decreased, which was attributed to water stress; the vegetation index (NDVI) then showed the highest correlations with the chlorophyll (a, b, and total) and carotene content of leaves [28].

Table 6. Carotenoid leaf content of the species.

Treatment	Leaf Carotenoid Content	
	Car $(mg \cdot m^{-2})$	Car $(mg \cdot g^{-1}; DW)$
T1	62.9 ± 3.2	7.2 ± 0.3
T2	75.8 ± 2.2	8.4 ± 0.4
T3	67.6 ± 4.1	7.9 ± 0.5

3.6. Effect of Irrigation Water Regimes on Soluble Carbohydrates Content

The irrigation water regimes had a slight effect on the soluble carbohydrate content on leaves of the species *T. tetragonioides*. The glucose and soluble carbohydrate content in leaves increased in the wet (T1) and dry (T3) treatments: glucose, 0.58 and 0.57 mg·mL^{-1}, respectively; soluble carbohydrates, 1.71 and 1.67 mg, respectively. These results are confirmed by Reference [29]. There was a decrease in glucose (0.54 mg·mL^{-1}) and soluble carbohydrate (1.59 g) content in leaves in the medium (T2) treatment (Table 7).

Table 7. Soluble carbohydrates content of leaves.

Treatment	Soluble Carbohydrates				
	Glucose $(mg \cdot mL^{-1})$	Area (cm^2)	Soluble Carbohydrates (mg)	DW (cm^2)	Soluble Carbohydrates (g)
T1	0.57 ± 0.05	3.42	1.67 ± 0.14	0.001	1.9 ± 0.15
T2	0.54 ± 0.03	3.42	1.59 ± 0.08	0.001	1.8 ± 0.09
T3	0.58 ± 0.03	3.42	1.71 ± 0.09	0.001	2.0 ± 0.12

3.7. Yield of Species

T. tetragonioides produced significant amounts of dry matter, which ranged from 82.7 to 66.1 g·plant^{-1}. The partition of the plant dry matter to plant organs was changed by the effect of the irrigation water regimes (Table 8). The fact that the species was irrigated during the vegetation period T1 (70%, wet treatment) significantly increased the dry biomass of the species at the harvest time, averaging 6616 kg·ha^{-1}. The dry matter of the species decreased when the soil water decreased in treatments T2 (50%, medium treatment) and T3 (30%, dry treatment). There was no significant difference between treatments. The obtained results confirmed that the species *T. tetragonioides* is tolerant to drought conditions. The yield of the crop shows that the drought had less effect than the salinity (6616–5288 kg DM·ha^{-1}). These results are confirmed by the previous study of Reference [3].

Table 8. Effect of irrigation water regimes on yield of the species. FW—fresh weight; DW—dry weight; FM—fresh matter.

Treatment	*Tetragonia tetragonioides*				
	FW (g·plant^{-1})	DW (g·plant^{-1})	Yield (%)	FM (kg·ha^{-1})	DM (kg·ha^{-1})
T1	1041.2 ± 12	82.7 ± 4	7.8 ± 0.3	83,284 ± 967	6609 ± 329
T2	913.7 ± 23	66 ± 3	7.2 ± 0.5	73,094 ± 1805	5289 ± 248
T3	966.2 ± 22	67.3 ± 3	6.8 ± 0.3	77,300 ± 1787	5377 ± 242

4. Conclusions

The experimental results showed several effects of the water irrigation regimes on the growth, mineral composition, and photosynthetic pigments of *T. tetragonioides*, as listed below.

- Plant growth (stem, leaves, and seeds) increased slightly with an enhancement of the water level (near the field capacity), and the growth difference between the drier water regimes was very low. This increase was probably due to the increase of stomatal conductance and, consequently, transpiration and CO_2 fixation were higher. Hence, it is not surprising that experimental results, in which the only variable was water application, agree quite well with this supposed theory.

- Leaf mineral composition of chloride and sodium are the main responsible ions for soil salinization and alkalization, respectively, in arid and semi-arid regions, enhanced by the decrease in soil water content. The content was very high in relation to other plants, showing its high capacity as a salt-removing species.

- There was a generally low decrease in phosphorus, calcium, potassium, iron, and magnesium in leaf content under drought conditions, probably due to the chloride and potassium competition.

- The total nitrogen leaf content of species showed very low variation, probably due to the same fertigation for all irrigation treatments.

-The minimum carotenoid amounts of chlorophyll a, chlorophyll b, and total chlorophyll were obtained from the wettest and the driest treatment in *T. tetragonioides* plants, probably due to higher plant senescence provoked by these regimes.

- The glucose and soluble carbohydrate contents of leaves increased in the driest treatments and had enhanced tolerance to drought conditions.

- The yield of the species increased in the wettest and the driest treatments.

In conclusion, it can be suggested that *T. tetragonioides* is a species tolerant to drought conditions. Its capacity as a halophyte and salt-removing species when the soil water content decreases was shown, suggesting its use in arid and semi-arid regions. Moreover, growth and yield differences in the various irrigation regimes were very low, which suggests another important advantage of these species—its cultivation under dry conditions, when used as a leafy vegetable for human consumption or for animal feeding. Nevertheless, more research is needed in order to test plant development under drier conditions in arid and semi-arid climates.

Author Contributions: The paper is the result of the collaboration among all authors; however, G.B. and J.B. contributed to the all sections. B.O. contributed to the sections on Effect of Irrigation Water Regimes on Mineral Composition in Plant Leaves.

Funding: This research received no external funding.

Conflicts of Interest: The authors declare no conflict of interest.

References

1. Costa, M.; Beltrao, J.; De Brito, J.C.; Guerrero, C. Turfgrass plant quality response to different water regimes. *WSEAS Trans. Environ. Dev.* **2011**, *7*, 167–176.

2. FAO. *The State of the World's Land and Water Resources for Food and Agriculture (SOLAW)—Managing Systems at Risk*; Food and Agriculture Organization of the United Nations and Earthscan: London, UK, 2011.

3. Bekmirzaev, G.; Beltrao, J.; Neves, M.A.; Costa, C. Climatical changes effects on the potential capacity of salt removing species. *Int. J. Geol.* **2011**, *5*, 79–85.

4. Szabolcs, I. Salt affected soils as the ecosystem for halophytes. In *Halophytes as a Resource for Livestock and for Rehabilitation of Degraded Lands*; Squires, V.R., Ayoub, A.T., Eds.; Kluwer Academic Publisher: London, UK, 1992; pp. 19–24.

5. Bartels, D.; Sunkar, R. Drought and salt tolerance in plants. *Crit. Rev. Plant. Sci.* **2005**, *24*, 23–58. [CrossRef]

6. Ansari, M.I.; Lin, T.P. Molecular analysis of dehydration in plants. *Int. Res. J. Plant. Sci.* **2010**, *1*, 21–25.

7. Boyer, J.S. Plant Productivity and Environment. *Science* **1982**, *218*, 443–448. [CrossRef] [PubMed]

8. Bray, E.A.; Bailey-Serres, J.; Weretilnyk, E. Responses to abiotic stresses. In *Biochemistry and Molecular Biology of Plants*; Buchanan, B.B., Gruissem, W., Jones, R.L., Eds.; American Society of Plant Physiologists: Rockville, MD, USA, 2000; pp. 1158–1203.

9. Ben-Asher, J.; Beltrao, J.; Costa, M.; Anaç, S.; Cuartero, J.; Soria, T. Modelling the effect of sea water intrusion on ground water salinity in agricultural areas in Israel, Portugal, Spain and Turkey. *Acta Hortic.* **2000**, *573*, 119–128. [CrossRef]

10. Nedjimi, B.; Daoud, Y.; Touati, M. Growth, water relations, proline and ion content of in vitro cultured *Atriplexhalimus* subsp. *schweinfurthii* as affected by CaCl$_2$. *Commun. Biom. Crop. Sci.* **2006**, *1*, 79–89.

11. Beltrao, J.; Correia, P.J.; Costa, M.; Gamito, P.; Santos, R.; Seita, J. The influence of nutrients on turfgrass response to treated wastewater application, under several saline conditions and irrigation regimes. *Environ. Proc.* **2014**, *1*, 105–113. [CrossRef]

12. FAO. Land and Plant Nutrition Management Service. Available online: http://www.fao.org/ag/agl/agll/spush/ (accessed on 15 May 2008).

13. Hamidov, A.; Khamidov, M.; Beltrão, J. Application of surface and groundwater to produce cotton in semi-arid Uzbekistan. *Asian Australas. J. Plant. Sci. Biotechnol.* **2013**, *7*, 67–71.

14. Yensen, N.P. Halophyte uses for the twenty-first century. In *Ecophysiology of High Salinity Tolerant Plants*; Springer: Dordrecht, The Netherlands, 2008; pp. 367–396.

15. Hasanuzzaman, M.; Nahar, K.; Alam, M.; Bhowmik, P.C.; Hossain, M.; Rahman, M.M.; Prasad, M.N.V.; Ozturk, M.; Fujita, M. Potential use of halophytes to remediate saline soils. *BioMed Res. Int.* **2014**. [CrossRef]

16. Cameron, R.W.F.; Harrison-Murray, R.S.; Atkinson, C.J.; Judd, H.L. Regulated deficit irrigation—A means to control growth in woody ornamentals. *J. Hortic. Sci. Biotechnol.* **2006**, *81*, 435–443. [CrossRef]

17. Grieve, C.M.; Suarez, D.L. Purslane (*Portulacaoleracea* L.): A halophytic crop for drainage water reuse systems. *Plant. Soil* **1997**, *192*, 277–283. [CrossRef]

18. Asher, J.B.; Beltrao, J.; Aksoy, U.; Anac, D.; Anac, S. Controlling and simulating the use of salt removing species. *Int. J. Energy Environ.* **2012**, *6*, 360–369.

19. Neves, A.; Miguel, M.G.; Marques, C.; Panagopoulos, T.; Beltrão, J. The combined effects of salts and calcium on growth and mineral accumulation of *Tetragoniate tragonioides*—A salt removing species. *WSEAS Trans. Environ. Dev.* **2008**, *4*, 1–5.

20. Veihmeyer, F.J.; Hendrickson, A.H. The moisture equivalent as a measure of the field capacity of soils. *Soil Sci.* **1931**, *32*, 181–193. [CrossRef]

21. Radojevic, M.; Bashkin, V.N. *Practical Environmental Analysis*; The Royal Society of Chemistry: Cambridge, UK, 1999.

22. Lichtenthaler, H.K. Chlorophylls and carotenoids: Pigments of photosynthetic biomembranes. *Meth. Enzymol.* **1987**, *148*, 350–382.

23. Dubois, M.; Gilles, K.A.; Hamilton, J.K.; Rebers, P.T.; Smith, F. Calorimetric method for determination of sugars and related substances. *Anal. Chem.* **1956**, *28*, 350–356. [CrossRef]

24. Pirzad, A.; Shakiba, M.R.; Zehtab-Salmasi, S.; Mohammadi, S.A.; Darvishzadeh, R.; Samadi, A. Effect of water stress on leaf relative water content, chlorophyll, proline and soluble carbohydrates in *Matricaria chamomilla* L. *J. Med. Plants Res.* **2011**, *5*, 2483–2488.

25. Bradford, K.J.; Hsiao, T.C. *Physiological Responses to Moderate Water Stress. Physiological Plant Ecology II*; Volume 12/B of the Series Encyclopedia of Plant Physiology; Springer-Verlag: Berlin, Germany, 1982; pp. 263–324.

26. Chartzoulakis, K.; Noitsakis, B.; Therios, I. Photosynthesis, plant growth and dry matter distribution in kiwifruit as influenced by water deficits. *Irrig. Sci.* **1993**, *14*, 1–5. [CrossRef]

27. Hussein, M.M.; Zaki, S.S. Influence of water stress onphotosynthesis pigments of some Fenugreek varieties. *J. Appl. Sci. Res.* **2013**, *9*, 5238–5245.

28. Köksal, E.S.; Üstün, H.; Özcan, H.; Güntürk, A. Estimating water stressed dwarf green bean pigment concentration through hyperspectral indices. *Pak. J. Bot.* **2010**, *42*, 1895–1901.

29. Redillas, M.C.; Park, S.H.; Lee, J.W.; Kim, Y.S.; Jeong, J.S.; Jung, H.; Bang, S.W.; Hahn, T.R.; Kim, J.K. Accumulation of trehalose increases soluble sugar contents in rice plants conferring tolerance to drought and salt tress. *Plant. Biotechnol. Rep.* **2012**, *6*, 89–96. [CrossRef]

agriculture

MDPI

Article

Soil Water Infiltration Model for Sprinkler Irrigation Control Strategy: A Case for Tea Plantation in Yangtze River Region

Yong-zong Lu [1,2,3], Peng-fei Liu [1], Aliasghar Montazar [2], Kyaw-Tha Paw U [3] and Yong-guang Hu [1,*]

1   Key Laboratory of Modern Agricultural Equipment and Technology, Ministry of Education Jiangsu Province, Jiangsu University, Zhenjiang 212013, China; luyongzong@126.com (Y.-z.L.); 18252585090@163.com (P.-f.L.)
2   Division of Agriculture and Natural Resources, UC Cooperative Extension, University of California, Imperial County, Holtville, CA 92250, USA; amontazar@ucanr.edu
3   Department of Land, Air and Water Resources, University of California, Davis, CA 95616, USA; ktpawu@ucdavis.edu
*   Correspondence: deerhu@ujs.edu.cn; Tel.: +86-138-1515-1176

Received: 23 July 2019; Accepted: 18 September 2019; Published: 20 September 2019

Abstract: The sprinkler irrigation method is widely applied in tea farms in the Yangtze River region, China, which is the most famous tea production area. Knowledge of the optimal irrigation time for the sprinkler irrigation system is vital for making the soil moisture range consistent with the root boundary to attain higher yield and water use efficiency. In this study, we investigated the characteristics of soil water infiltration and redistribution under the irrigation water applications rates of 4 mm/h, 6 mm/h, and 8 mm/h, and the slope gradients of 0°, 5°, and 15°. A new soil water infiltration model was established based on water application rate and slope gradient. Infiltration experimental results showed that soil water infiltration rate increased with the application rate when the slope gradient remained constant. Meanwhile, it decreased with the increase in slope gradient at a constant water application rate. In the process of water redistribution, the increment of volumetric water content (VWC) increased at a depth of 10 cm as the water application rate increased, which affected the ultimate infiltration depth. When the slope gradient was constant, a lower water application rate extended the irrigation time, but increased the ultimate infiltration depth. At a constant water application rate, the infiltration depth increased with the increase in slope gradient. As the results showed in the infiltration model validation experiments, the infiltration depths measured were 38.8 cm and 41.1 cm. The relative errors between measured infiltration depth and expected value were 3.1% and 2.7%, respectively, which met the requirement of the soil moisture range consistent with the root boundary. Therefore, this model could be used to determine the optimal irrigation time for developing a sprinkler irrigation control strategy for tea fields in the Yangtze River region.

Keywords: water application rate; slope gradient; infiltration depth; optimal irrigation time

1. Introduction

Tea (*Camellia sinensis*) is a subtropical plant, which grows well in a warm and wet climate under an optimal growth temperature around 20 °C with annual rainfall of 1500 mm. The Yangtze River region is the most famous tea production area. Climate abnormality in recent decades caused an uneven spatial distribution of rainfall in this area. Inadequate rainfall cannot provide enough water for tea growth, which seriously affects both the yield and quality of the tea plants [1,2]. Under water stress, the photosynthetic and respiration rates of tea leaves decrease [3], as well as their chlorophyll content, the water content of shoots, the root activity, and the root weight per unit volume [4]. Meanwhile, quality components such as amino acids, caffeine, and water extracts are also reduced, resulting in the deterioration of the tea quality [5–8].

The sprinkler irrigation method is widely applied to save water and counter drought with great improvement with regard to water use efficiency, irrigation uniformity, labor saving, and crop yields [9]. The traditional control strategy for sprinkler irrigation is usually based on the upper limit and lower limit of soil moisture required for the growth of certain plants [10–12]. This kind of control strategy partly provides the required water for tea plant growth and saves some applied water compared to manual irrigation. However, it leads to the ultimate infiltration depth exceeding the boundary of the tea plant root system, resulting in a waste of water resources. Thus, it is necessary to improve the water use efficiency (WUE) of sprinkler irrigation systems, especially for tea fields in the Yangtze River region.

Infiltration is the process of water entering the soil. The main goal of operating an irrigation system is to apply the required infiltration depth for specific plants with high WUE [3,7,9,13,14]. Normally, infiltration rate is determined by measuring, modeling, and predicting the surface runoff [15–17]. To avoid low-WUE problems, it is necessary to have a good understanding of soil infiltration characteristics. Infiltration rate variation results from many causes such as soil property, water application rate, and terrain [18–20]. Infiltration theories and models were developed by several researchers, including the Green–Ampt model, Kostiakov model, modified Kostiakov model, and Smith and Parlange model [21–23]. The suitability of an infiltration model for a particular region is subject to soil type and field conditions. Different infiltration models were applied to certain soil types and certain site conditions [24]. Over the years, several comparative analyses of various infiltration models were conducted to assess the suitability of various models for different soil types under varying field conditions to estimate the infiltration rates and infiltration potentials of soil [25]. Feng, Deng, Zhang, and Guo [26] pointed out that the cumulative infiltration depth firstly decreased then increased with the increase in slope gradient, and the turning point was the threshold of the slope gradient. Recent researches focused on the effects of slope gradient and water application rate on soil water infiltration and redistribution without any plants; however, they failed to reveal the situation with plants [27,28]. To some extent, the morphological, quantitative, depth, type, and distribution characteristics of plant roots affect soil water moisture and its redistribution [29–32]. The infiltration characteristics of soil change with the root length density and root surface area density [33,34]. With the increase in root volume and dry weight of root, the infiltration rate shows an increasing trend [35]. Soil moisture content is lower in soil layers with denser root density, while water content in soil layers without roots is significantly increased [36]. If the moisture content range is not consistent with the root range of the tea plant, it will result in inefficient irrigation or overirrigation. Therefore, an optimal irrigation time and infiltration depth are vital for tea plants in the context of a water-saving irrigation control strategy.

Tea farms in the middle and lower Yangtze River regions are located in a hilly area, with a slope gradient generally less than 15° [27]. Based on the topographic features of tea farms, the specific goals of this study were to (1) investigate the effects of slope gradient and water application rate on soil water infiltration and redistribution, and (2) provide a new infiltration model for determining the optimal stopping time for a tea plantation sprinkler irrigation control system. The model is based on tea root length, water application rate, and slope gradient, and results in the infiltration depth being consistent with tea plant's root system, which validates the precision control for tea plant irrigation.

2. Materials and Methods

2.1. Materials

This study was conducted at a tea farm located in the middle–lower Yangtze River region, east China, which has a moderate sub-tropical climate with a mean annual precipitation of 1029.1 mm and an average annual temperature of 15.5 °C. The annual reference evapotranspiration (ET_0) was 892.24 mm, which was observed over the last 55 years (1961–2015) [37]. The topography of the experimental site is a hilly ground with an average altitude of 18.5 m (latitude 32°01′35″ north (N), longitude 119°40′21″ east (E)).

As shown in Figure 1, the box frame sprinkler irrigation system was composed of a box frame, pump station, main pipe, distribution pipe, lateral pipe, standpipe, and sprinkler. The distance between two sprinklers was 9.0 m. In order to reduce the pressure difference between sprinklers and improve the uniformity of spraying water, lateral pipes with a length of 4.5 m were laid along the slope. Four sprinklers were set up around the box frame with an effective radius of 7.0 m. The box frame was 4.5 m in length, 0.8 m in width, and 0.8 m in depth, and the adjustable range of the gradient is 0–15°. To ensure the box frame was well drained, drain holes (diameter 1.0 cm) were set up at the bottom of the box. The exterior of the box was made of an acrylic plate (80.0 cm × 0.5 cm, transparency 99.0%) to facilitate the measurement of infiltration depth. The sampled tea cultivar was Anji white tea, which was five years old. Six tea plants were planted in the frame box with distances of 0.35, 1.05, 1.75, 2.75, 3.45, and 4.15 m (Figure 2). Three kinds of plastic impact-driven sprinklers were selected. Technical parameters of the sprinklers are listed in Table 1. Three kinds of water application rates were set up using the different types of sprinklers with flow rates of 0.4, 0.6, and 0.8 m^3·h^{-1}. A plastic shade was set up above the experiment area on rainy days to avoid the influence of rainfall.

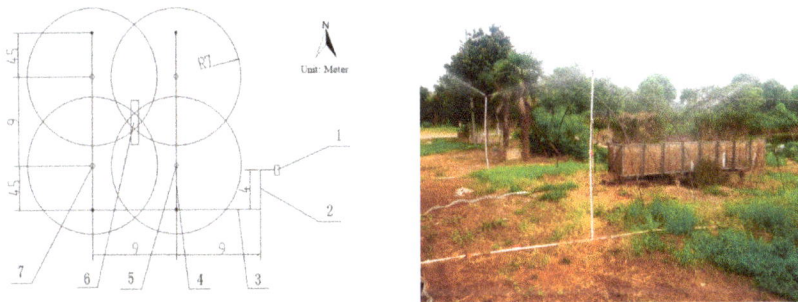

Figure 1. Box frame sprinkler irrigation system: 1—pump station; 2—main pipe; 3—distribution pipe; 4—lateral pipe; 5—standpipe; 6—box frame; 7—sprinkler.

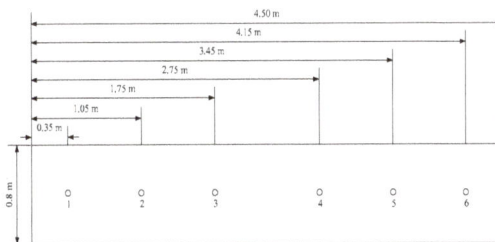

Figure 2. Tea plant distribution in the box frame.

Table 1. Technical parameters of the sprinkler.

No.	Operating Pressure (MPa)	Nozzle Diameter (mm)	Flow Rate (m^3·h^{-1})	Pattern Radius (m)	Rotation Cycle (s)
1	0.3	3.0	0.4	7.0	18.0
2	0.3	3.5	0.6	7.0	18.0
3	0.3	4.0	0.8	7.0	18.0

The classification of soil texture was based on the World Reference Base (WRB) soil classification system [38]. The soil was collected from the experimental site at depths of 0–10.0 cm, 10.0–20.0 cm, 20.0–30.0 cm, 30.0–40.0 cm, 40.0–50.0 cm, and 50.0–80.0 cm. Bulk density and saturated VWC were measured using the oven dry method. A laser diffraction particle size analyzer (Mastersizer 3000, Malvern Panalytical, UK) was used to measure the particle composition. The soil texture at a depth of 0–40 cm was sandy loam, and that under 40 cm was loam (Table 2). A soil moisture sensor (TRIME PICO 64, IMKO, Ettlingen, Germany) was used to measure VWC, and the measuring accuracy was ± 1%.

Table 2. Physical properties and the particle composition of the soil for experiment. VWC—volumetric water content.

Sampling Depth (cm)	Size Composition (%)			Bulk Density (g·cm^{-3})	Saturated VWC (%)	Soil Texture
	<0.002 mm	0.002–0.02 mm	0.02–2 mm			
0–10.0	0	37.6	62.4	1.2	47.0	
10.0–20.0	0	29.3	70.7	1.4	47.0	
20.0–30.0	0	39.7	60.3	1.4	49.0	Sandy loam
30.0–40.0	0	26.8	73.2	1.5	48.0	
40.0–50.0	4.4	42.8	52.8	1.7	44.0	
50.0–80.0	8.6	40.6	50.8	1.6	43.0	Loam

2.2. Methods

2.2.1. Sensor Layout

The soil in the experimental box frame was taken from the sprinkler irrigation area in the experimental tea farm. The sampling depths were 0–10.0 cm, 10.0–20.0 cm, 20.0–30.0 cm, 30.0–40.0 cm, 40.0–50.0 cm, and 50.0–80.0 cm (Figure 3). The sampled soil was air-dried, ground, and screened before layering it into the experimental box frame. The surface of the filled soil was hacked to reduce the influence of artificial compaction on soil water infiltration.

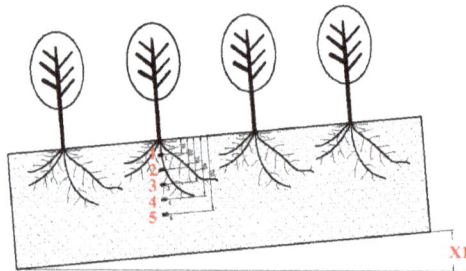

Figure 3. Sensor layout. TRIME PICO 64 soil moisture sensors were arranged along the roots of tea plants. Numbers 1–5 represent the sensors located at depths of 10.0 cm, 20.0 cm, 30.0 cm, 40.0 cm, and 50.0 cm, respectively. X1 represents the slope gradient, which could be adjusted from 0–15°.

Infiltration depth and VWC were measured in this experiment. All soil moisture sensors were calibrated using the oven dry method before set-up. The time interval of data acquisition was 1.0 min.

Soil type in the experimental site was homogeneous sandy loam. The location of the color gradient of the soil was marked, and the depth of position was measured as the infiltration depth. The water infiltration process started with the irrigation and stopped when the irrigation stopped, and then the water continued to infiltrate. When the VWC of each layer no longer increased and showed a decreasing trend, the process of water redistribution stopped.

2.2.2. Characteristics of Soil Water Infiltration

Characteristics of the soil water infiltration experiments were determined in the frame box from 1–26 August 2016 (Table 3). Nine treatments were selected with three typical kinds of slope gradients and water application rates. The irrigation was stopped when the infiltration depth was 20.0 cm due to the average length of the tea plant roots. SPSS 17.0 was used to conduct the multivariate regression analysis of the relationship between slope gradient, water application rate, and the ratio of stopping irrigation depth to infiltration depth. The statistical tools root-mean-squared error (*RMSE*) and coefficient of determination (R^2) were employed to validate the accuracy of the models built in this study. The calculations are presented below.

Table 3. Sprinkler irrigation schedule.

Treatment	Slope Gradient (°)	Water Application Rate (mm·h⁻¹)
T1	0	4.0
T2	0	6.0
T3	0	8.0
T4	5.0	4.0
T5	5.0	6.0
T6	5.0	8.0
T7	15.0	4.0
T8	15.0	6.0
T9	15.0	8.0

2.2.3. Infiltration Model Validation Experiments

The soil water infiltration model was established based on tea root depth, water application rate, and slope gradient. The reliability of the model was evaluated in the box frame experiments based on two cases which represented the most common terrains of tea farms in the Yangtze River region: (1) gradient 0° and 8.0 mm·h^{-1} for water application rate; (2) gradient 8° and 4.0 mm·h^{-1} for water application rate. The required irrigation time and expected infiltration depth were calculated. The VWC at depths of 10.0 cm, 20.0 cm, 30.0 cm, and 40.0 cm was measured. The ultimate infiltration depth was compared with the expected value to obtain the relative error between the measured and expected value.

3. Results and Discussion

3.1. VWC at Different Slope Gradients and Water Application Rates

As the sprinkler irrigation stopped, under the action of gravity potential and matric potential, water redistribution began. After 24 h, the VWC no longer increased and showed a downward trend, signifying that the process of water redistribution stopped (Figure 4).

Compared with the VWC before irrigation, VWC at 20.0 cm increased at the time of stopping irrigation, and the increment of VWC at 10 cm under different treatments was different. When the slope gradient was 0°, the increments of VWC at 10.0 cm were 11.1%, 10.0%, and 4.5%, respectively. When the slope was 5°, the increments were 6.1%, 9.1%, and 8.0%, respectively. When the slope was 15°, the increments were 9.0%, 5.9%, and 4.0%, respectively (Figure 4b).

Twelve hours after irrigation, soil water at the soil layer depth of 0–20.0 cm started to redistribute (Figure 4c). Compared with VWC at the time of stopping irrigation, VWC at 10.0 cm decreased by 1.1% for the T2 treatment. For the other eight treatments, the VWC at 10.0 cm increased by 2.3%, 5.0%, 9.7%, 4.1%, 4.3%, 1.3%, 4.3%, and 6.5%, respectively. VWC at 20.0 cm increased for all nine treatments, whereas VWC at 30 cm, 40 cm, and 50 cm showed no change.

Infiltration depth showed an increasing trend under the processing of water redistribution (Figure 4d). Compared with VWC at 12 h after the irrigation, VWC at 10.0 cm showed decreasing trends for all nine treatments, while VWC at 20.0 cm continued increasing. This was probably because the amount of water drawn from the upper soil at 10.0 cm was less than the water absorbed by the lower soil layer, which resulted in a continuous decrease in VWC at 10.0 cm and a sustained increase in VWC below 20.0 cm.

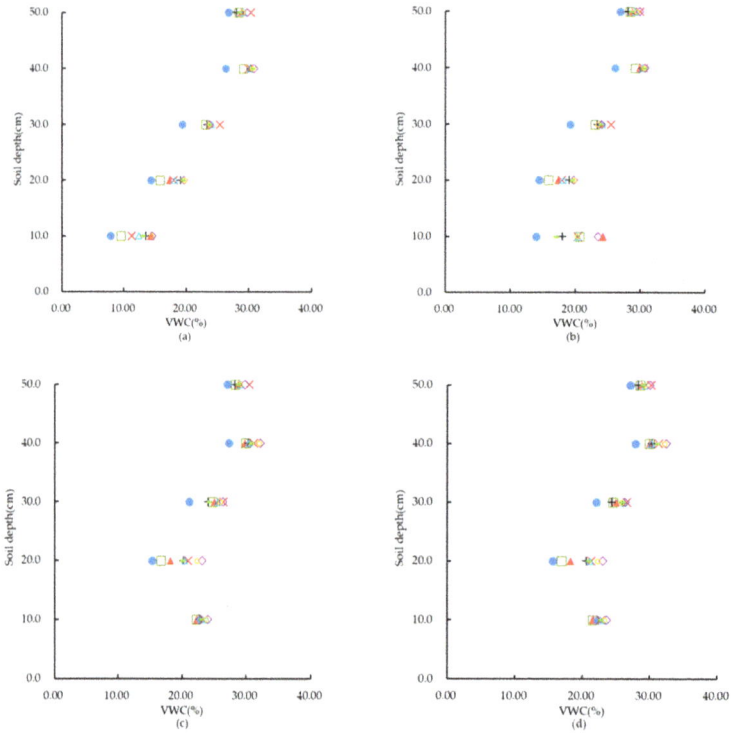

Figure 4. Volumetric water content (VWC) before irrigation, at the time of stopping irrigation, and 12 h and 24 h after sprinkler irrigation with different treatments, T1(□), T2(▲), T3(+), T4(●), T5(×), T6( ), T7(◇), T8( ), and T9(–): (**a**) before irrigation; (**b**) at the time of stopping irrigation; (**c**) 12 h after irrigation; (**d**) 24 h after irrigation.

3.2. *Effect of Water Application Rate and Slope Gradient on the Infiltration Depth and Rate*

When the water application rate was constant, the infiltration depth increased as the slope gradient increased (Figure 5). This is because the pressure of the water perpendicular to the direction of the slope decreased, which increased the infiltration depth (Table 4).

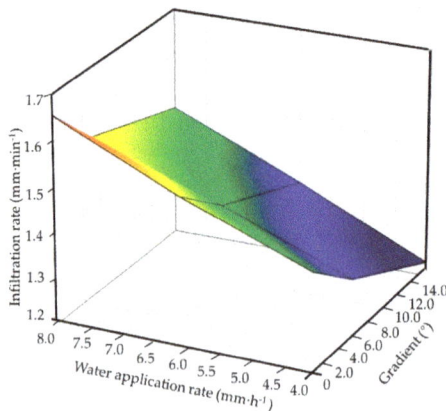

Figure 5. Infiltration rate under various water application rates and slope gradients.

Table 4. The ratios of different slope gradients and water application rates.

Treatment	Slope Gradient (°)	Water Application Rate (mm·h⁻¹)	Infiltration Depth (cm)	Ratio of Stopping Irrigation Depth to Infiltration Depth (%)
T1	0	4.0	40.9	48.9
T2	0	6.0	34.9	57.4
T3	0	8.0	32.5	61.6
T4	5.0	4.0	42.6	46.9
T5	5.0	6.0	41.3	48.4
T6	5.0	8.0	35.8	55.9
T7	15.0	4.0	43.2	46.3
T8	15.0	6.0	42.5	47.0
T9	15.0	8.0	37.3	53.6

Through the multivariate regression analysis, it could be concluded that the relationship between slope gradient, water application rate, and the ratio of stopping irrigation depth to infiltration depth was as follows:

$$Y_1 = -0.004X_1 + 0.024X_2 + 0.401, \tag{1}$$

where Y_1 is the ratio of stopping irrigation depth to infiltration depth (mm), X_1 is the slope gradient (°), and X_2 is the water application rate (mm·h⁻¹). The R^2 was 0.83 and the *RMSE* was 0.02 mm.

Using multivariate regression analysis, the linear regression equations for infiltration rate, slope gradient, and water application rate were obtained. The R^2 was 0.92 and the *RMSE* was 0.02 mm·h⁻¹. The linear regression equations were as follows:

$$Y_2 = -0.012X_1 + 0.058X_2 + 1.181, \tag{2}$$

$$L_1 = L \times (-0.04X_1 + 0.24X_2 + 4.01), \tag{3}$$

$$T = \frac{L_1}{Y_2} = \frac{(-0.04X_1 + 0.24X_2 + 4.01)L}{-0.012X_1 + 0.058X_2 + 1.181}, \tag{4}$$

where Y_2 is the infiltration rate (mm·min⁻¹), L_1 is the required infiltration depth (cm), L is the root depth of the tested tea plant (cm), and T is the required sprinkler irrigation time (min).

3.3. The Reliability of Infiltration Model Testing

The average root length of the tested tea plants was 40.0 cm. When the slope gradient was 0° and the water application rate was 8.0 mm·h⁻¹, the required irrigation time was 144 min. The expected stopping irrigation depth and observed infiltration depth were 23.7 cm and 40.0 cm, respectively.

Before irrigation, the VWC of each layer was 11.4%, 18.1%, 22.1%, 28.9%, and 26.7%, respectively (Figure 6). After 63 min of irrigation, the VWC at 10 cm was 11.6%; then, it increased gradually. Additionally, 125 min after irrigation, the VWC for the 10-cm and 20-cm soil layers was 15.8%, 18.1%, respectively, after which it increased gradually. The system stopped after 144 min of irrigation. At this time, the VWC of each soil layer was 19.2%, 18.3%, 22.0%, 28.9%, and 26.6%, and the infiltration depth was 23.2 cm. Then, 24 h after irrigation, the water infiltration stopped. The VWC of each layer was 20.8%, 19.0%, 23.2%, 29.1%, and 26.7%, and the infiltration depth measured was 38.75 cm. Compared with the required irrigation time and infiltration depth, the measured infiltration depth was 23.2 cm, giving an error between the measured and required value of 2.0%. The measured ultimate infiltration depth was 38.8 cm, giving an error between the measured and expected value of 3.1%.

When the slope gradient was 8° and the water application rate was 4.0 mm·h⁻¹, the required irrigation time was 141 mins, and the expected infiltration depth and observed infiltration depth were 18.6 cm and 40.0 cm respectively. Before irrigation, the VWC of each layer was 15.0%, 19.1%, 23.1%, 28.5%, and 26.1%, respectively (Figure 7). After 141 min of irrigation, the sprinkler system stopped. The VWC of each layer was 21.5%, 19.2%, 23.1%, 28.5%, and 26.2%, respectively, and the infiltration depth was 18.7 cm. Then, 24 h after irrigation, the water infiltration stopped. The VWC of each layer was 20.0%, 21.1%, 24.7%, 28.7%, and 26.2%, and the ultimate infiltration depth measured was 41.1 cm. The measured infiltration depth was 18.9 cm, and the relative error between the measured and expected value was 1.4%. The measured ultimate infiltration depth was 41.1 cm, and the relative error between the measured and expected value was 2.7%.

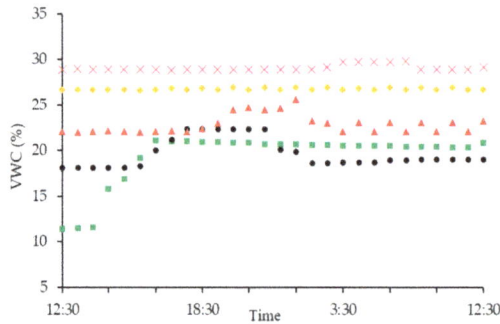

Figure 6. VWC at the slope gradient of 0° and the water application rate of 8.0 mm·h^{-1}, with soil depths of 10 cm (■), 20 cm (●) 30 cm (▲), 40 cm (×), and 50 cm (♦).

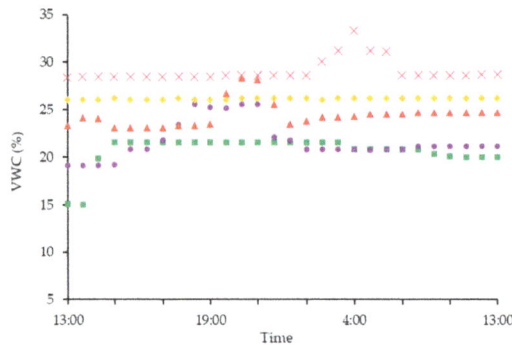

Figure 7. VWC at the slope gradient of 8° and the water application rate of 4.0 mm·h^{-1}, with soil depths of 10 cm (■), 20 cm (●) 30 cm (▲), 40 cm (×), and 50 cm (♦).

4. Discussion

With the increase in slope gradient, the required water application rate decreased (Figure 4). When the water application rate was close to the required water application rate, the soil pores were gradually filled with water. This reduced the infiltration capacity and led to a decrease in the increment of VWC.

For a constant water application rate, the infiltration rate decreased as the slope gradient increased. This is because a larger water application rate caused a larger kinetic energy of the sprayed water droplets [39]. Therefore, as the infiltration rate and water application rate increased, the water pressure of the soil surface and soil layers all increased over time. As the kinetic energy of water droplets increased, more pressure was applied on the infiltration water, which resulted in an acceleration of the infiltration rate. The infiltration rate also showed a decreasing trend with the increase in water application rate. The reason was that the component force of the same thickness of the aquifer along the slope direction increased with an increase in slope, and the pressure perpendicular to the slope direction was reduced. Therefore, the pressure of infiltration water reduced, as did the infiltration rate [20,40–42].

At a constant slope gradient, the larger water application rate led to a lower infiltration depth. This was caused by the water application rate being lower than its own infiltration capacity, resulting in the water continuing to infiltrate over time [41]. With the increase in water application rate, the kinetic energy of the water droplets sprayed from the nozzle increased correspondingly, resulting in a reduction in infiltration capacity. A lower water application rate extended the irrigation time, but it was conducive to the vertical movement of water, which deepened the infiltration depth.

As we all know, the infiltration characteristic has a strong relationship with the kinetic energy of water drops and the change in physical properties of the soil surface, such as soil type, vegetation type, and terrain [43–45]. We know our research is limited; however, the results were obvious in the two selected experiments, where the ultimate infiltration was consistent with the boundary of the tea plant root system. Based on the characteristics of irrigation water infiltration and redistribution, the new infiltration model can be used to determine the required irrigation time for developing a sprinkler irrigation control strategy.

5. Conclusions

The sprinkler irrigation method is widely used for tea plants in the Yangtze River region. The traditional control strategy for sprinkler irrigation is based on the upper limit and lower limit of the required soil moisture. However, this strategy always causes the ultimate infiltration depth to exceed the boundary of the tea plant root system, leading to a waste of water. In this study, a new soil water infiltration model was provided by investigating the characteristics of soil water infiltration and redistribution under different water application rates and gradient slopes.

The infiltration characteristics showed that the infiltration rate changed with the water application rate and slope gradient. Water redistribution processes showed that the increment of VWC at 10.0 cm was different for various combinations of water application rate and slope gradient. Those differences affected the ultimate infiltration depth of the soil. When the slope gradient was kept constant, a lower water application rate led to a longer irrigation time, but it increased the ultimate infiltration depth. When the water application rate was kept constant, the infiltration depth increased with the increase in slope gradient. Based on the new soil water infiltration model, the ultimate infiltration depth was basically consistent with the boundary of the tea plant roots. Therefore, the model established in this paper can be generally applied to automatic sprinkler irrigation systems for tea fields in the Yangtze River region.

Author Contributions: Y.-g.H. conceptualized and designed the experiments. Y.-z.L. and P.-f.L. performed the experiments. Y.-z.L. analyzed the data and wrote the manuscript. K.-T.P.U. and A.M. contributed significant comments to improve the quality and language of the manuscript.

Funding: This research was funded by Jiangsu Agriculture Science and Technology Innovation Fund (CX(16)1045), the Key R&D Programs of Jiangsu Province and Zhenjiang (BE2016354, NY20160120037), the Project of Postgraduate Innovation of Jiangsu Province (KYCX17-1788), the China and Jiangsu Postdoctoral Science Foundation (2016M600376, 1601032C), the Six Talent Peaks Program in Jiangsu Province (2015-ZBZZ-021), the Priority Academic Program Development of Jiangsu Higher Education Institutions (2014-37).

Acknowledgments: The authors are grateful for the financial support from the Jiangsu Agriculture Science and Technology Innovation Fund (CX(16)1045), the Key R&D Programs of Jiangsu Province and Zhenjiang (BE2016354, NY20160120037), the Project of Postgraduate Innovation of Jiangsu Province (KYCX17-1788), the China and Jiangsu Postdoctoral Science Foundation (2016M600376, 1601032C), the Six Talent Peaks Program in Jiangsu Province (2015-ZBZZ-021), the Priority Academic Program Development of Jiangsu Higher Education Institutions (2014-37), and the China Scholarship Council (201708320220).

Conflicts of Interest: The authors declare no conflict of interest.

References

1.   Ding, Y.; Wang, W.; Song, R.; Shao, Q.; Jiao, X.; Xing, W. Modeling spatial and temporal variability of the impact of climate change on rice irrigation water requirements in the middle and lower reaches of the Yangtze River, China. *Agric. Water Manag.* **2017**, *193*, 89–101. [CrossRef]

2. Song, S.; Xu, Y.P.; Zhang, J.X.; Li, G.; Wang, Y.F. The long-term water level dynamics during urbanization in plain catchment in Yangtze River Delta. *Agric. Water Manag.* **2016**, *174*, 93–102. [CrossRef]

3. Maritim, T.K.; Kamunya, S.M.; Mireji, P.; Mwendia, C.; Muoki, R.C.; Cheruiyot, E.K.; Wachira, F.N. Physiological and biochemical response of tea [*Camellia sinensis* (L.) O. Kuntze] to water-deficit stress. *J. Hortic. Sci. Biotechnol.* **2015**, *90*, 395–400. [CrossRef]

4. Netto, L.A.; Jayaram, K.M.; Puthur, J.T. Clonal variation of tea [*Camellia sinensis* (L.) O. Kuntze] in countering water deficiency. *Physiol. Mol. Biol. Plants Int. J. Funct. Plant Biol.* **2010**, *16*, 359–367. [CrossRef] [PubMed]

5. Chakraborty, U.; Dutta, S.; Chakraborty, B. Drought induced biochemical changes in young tea leaves. *Indian J. Plant Physiol. India* **2001**, *6*, 103–106.

6. Kigalu, J.M.; Kimambo, E.I.; Msite, I.; Gembe, M. Drip irrigation of tea (*Camellia sinensis* L.): 1. Yield and crop water productivity responses to irrigation. *Agric. Water Manag.* **2008**, *95*, 1253–1260. [CrossRef]

7. Kumar, R.; Bisen, J.S.; Choubey, M.; Singh, M.; Bera, B. Influence of Changes Weather Conditions on Physiological and Biochemical Characteristics of Darjeeling Tea (*Camellia sinensis* L.). *Glob. J. Biol. Agric. Health. Sci.* **2016**, *5*, 55–60.

8. Lin, S.K.; Lin, J.; Liu, Q.L.; Ai, Y.F.; Ke, Y.Q.; Chen, C.; Zhang, Z.Y.; He, H. Time-course of photosynthesis and non-structural carbon compounds in the leaves of tea plants (*Camellia sinensis* L.) in response to deficit irrigation. *Agric. Water Manag.* **2014**, *144*, 98–106. [CrossRef]

9. Saretta, E.; de Camargo, A.P.; Botrel, T.A.; Frizzone, J.A.; Koech, R.; Molle, B. Test methods for characterising the water distribution from irrigation sprinklers: Design, evaluation and uncertainty analysis of an automated system. *Biosyst. Eng.* **2018**, *169*, 42–56. [CrossRef]

10. Ratliff, L.F.; Ritchie, J.T.; Cassel, D.K. Field-Measured Limits of Soil Water Availability as Related to Laboratory-Measured Properties 1. *Soil Sci. Soc. Am. J.* **1983**, *47*, 770–775. [CrossRef]

11. Thompson, R.B.; Gallardo, M.; Valdez, L.C.; Fernández, M.D. Using plant water status to define threshold values for irrigation management of vegetable crops using soil moisture sensors. *Agric. Water Manag.* **2007**, *88*, 147–158. [CrossRef]

12. Wei, Y.; Wang, Z.; Wang, T.; Liu, K. Design of real time soil moisture monitoring and precision irrigation systems. *Trans. Chin. Soc. Agric. Eng.* **2013**, *29*, 80–86.

13. Liu, Y.-Y.; Wang, A.-Y.; An, Y.-N.; Lian, P.-Y.; Wu, D.-D.; Zhu, J.-J.; Meinzer, F.C.; Hao, G.-Y. Hydraulics play an important role in causing low growth rate and dieback of aging *Pinus sylvestris* var. mongolica trees in plantations of Northeast China. *Plant Cell Environ.* **2018**, *41*, 1500–1511. [CrossRef] [PubMed]

14. Vaz, C.M.P.; Jones, S.; Meding, M.; Tuller, M. Evaluation of Standard Calibration Functions for Eight Electromagnetic Soil Moisture Sensors. *Vadose Zone J.* **2013**, *12*. [CrossRef]

15. Al-Ghobari, H.M.; El-Marazky, M.S.; Dewidar, A.Z.; Mattar, M.A. Prediction of wind drift and evaporation losses from sprinkler irrigation using neural network and multiple regression techniques. *Agric. Water Manag.* **2018**, *195*, 211–221. [CrossRef]

16. AL-Kayssi, A.W.; Mustafa, S.H. Modeling gypsifereous soil infiltration rate under different sprinkler application rates and successive irrigation events. *Agric. Water Manag.* **2016**, *163*, 66–74. [CrossRef]

17. Diamond, J.; Shanley, T. Infiltration rate assessment of some major soils. *Ir. Geogr.* **2003**, *36*, 32–46. [CrossRef]

18. Fu, B.; Wang, Y.; Zhu, B.; Wang, D.; Wang, X.; Wang, Y.; Ren, Y. Experimental study on rainfall infiltration in sloping farmland of purple soil. *Trans. Chin. Soc. Agric. Eng.* **2008**, *2008*. [CrossRef]

19. Liu, Z.; Li, P.; Hu, Y.; Wang, J. Wetting patterns and water distributions in cultivation media under drip irrigation. *Comput. Electron. Agric.* **2015**, *112*, 200–208. [CrossRef]

20. Mu, W.; Yu, F.; Li, C.; Xie, Y.; Tian, J.; Liu, J.; Zhao, N.; Mu, W.; Yu, F.; Li, C.; et al. Effects of Rainfall Intensity and Slope Gradient on Runoff and Soil Moisture Content on Different Growing Stages of Spring Maize. *Water* **2015**, *7*, 2990–3008. [CrossRef]

21. Zakwan, M.; Muzzammil, M.; Alam, J. Application of spreadsheet to estimate infiltration parameters. *Perspect. Sci.* **2016**, *8*, 702–704. [CrossRef]

22. Smith, R.E.; Parlange, J.-Y. A parameter-efficient hydrologic infiltration model. *Water Resour. Res.* **1978**, *14*, 533–538. [CrossRef]

23. Parhi, P.K.; Mishra, S.K.; Singh, R. A Modification to Kostiakov and Modified Kostiakov Infiltration Models. *Water Resour. Manag.* **2007**, *21*, 1973–1989. [CrossRef]

24. Mazloom, H.; Foladmand, H. Evaluation and determination of the coefficients of infiltration models in Marvdasht region, Fars province. *Int. J. Adv. Biol. Biomed. Res.* **2013**, *1*, 822–829.

25. Wilson, R.L. Comparing Infiltration Models to Estimate Infiltration Potential at Henry V Events. Bachelor's Thesis, Portland State University, Portland, OR, USA, 2017.

26. Feng, H.; Deng, L.S.; Zhang, C.L.; Guo, Y.B. Effect of Ground Slope on Water Infiltration of Drip Irrigation. *J. Irrig. Drain.* **2010**, *29*, 14–15.

27. Pengfei, L.; Yongguang, H.; Feng, J.; Sheng, W. Influence of sloping tea fields on soil moisture migration. *IFAC Pap.* **2018**, *51*, 565–569. [CrossRef]

28. Zhao, P.; Shao, M.; Melegy, A.A. Soil Water Distribution and Movement in Layered Soils of a Dam Farmland. *Water Resour. Manag.* **2010**, *24*, 3871–3883. [CrossRef]

29. Jha, S.K.; Gao, Y.; Liu, H.; Huang, Z.; Wang, G.; Liang, Y.; Duan, A. Root development and water uptake in winter wheat under different irrigation methods and scheduling for North China. *Agric. Water Manag.* **2017**, *182*, 139–150. [CrossRef]

30. Yan, Y.; Dai, Q.; Yuan, Y.; Peng, X.; Zhao, L.; Yang, J. Effects of rainfall intensity on runoff and sediment yields on bare slopes in a karst area, SW China. *Geoderma* **2018**, *330*, 30–40. [CrossRef]

31. Wang, Y.; Zhang, X.; Chen, J.; Chen, A.; Wang, L.; Guo, X.; Niu, Y.; Liu, S.; Mi, G.; Gao, Q. Reducing basal nitrogen rate to improve maize seedling growth, water and nitrogen use efficiencies under drought stress by optimizing root morphology and distribution. *Agric. Water Manag.* **2019**, *212*, 328–337. [CrossRef]

32. Wang, X.; Zhou, Y.; Wang, Y.; Wei, X.; Guo, X.; Zhu, D. Soil water characteristic of a dense jujube plantation in the semi-arid hilly Regions of the Loess Plateau in China. *J. Hydraul. Eng.* **2015**, *46*, 263–270.

33. Li, N.; Kang, Y.; Li, X.; Wan, S.; Xu, J. Effect of the micro-sprinkler irrigation method with treated effluent on soil physical and chemical properties in sea reclamation land. *Agric. Water Manag.* **2019**, *213*, 222–230. [CrossRef]

34. Li, Z.; Xu, X.; Pan, G.; Smith, P.; Cheng, K. Irrigation regime affected SOC content rather than plow layer thickness of rice paddies: A county level survey from a river basin in lower Yangtze valley, China. *Agric. Water Manag.* **2016**, *172*, 31–39. [CrossRef]

35. Zhang, F.; Niu, X.; Zhang, Y.; Xie, R.; Liu, X.; Li, S.; Gao, S. Studies on the Root Characteristics of Maize Varieties of Different Eras. *J. Integr. Agric.* **2013**, *12*, 426–435. [CrossRef]

36. Liu, X.; Wang, Y.; Ma, L.; Liang, Y. Relationship between Deep Soil Water Vertical Variation and Root Distribution in Dense Jujube Plantation. *Trans. Chin. Soc. Agric. Mach.* **2013**, *44*, 90–97.

37. Chu, R.; Li, M.; Shen, S.; Islam, A.R.M.d.T.; Cao, W.; Tao, S.; Gao, P. Changes in Reference Evapotranspiration and Its Contributing Factors in Jiangsu, a Major Economic and Agricultural Province of Eastern China. *Water* **2017**, *9*, 486. [CrossRef]

38. FAO. *World Reference Base for Soil Resources 2014: International Soil Classification System for Naming Soils and Creating Legends for Soil Maps*; FAO: Rome, Italy, 2014.

39. Cui, S.F.; Pan, Y.H.; Wu, Q.Y.; Zhang, Z.H.; Zhang, B.X. Simulation of Runoff for Varying Mulch Coverage on a Sloped Surface. *Appl. Mech. Mater.* **2013**, *409*, 339–343. [CrossRef]

40. Marshall, S.J. Hydrology. In *Reference Module in Earth Systems and Environmental Sciences*; Elsevier: Amsterdam, The Netherlands, 2013.

41. Ge, S.; Gorelick, S.M. Hydrology, Floods and Droughts|Groundwater and Surface Water. In *Encyclopedia of Atmospheric Sciences*; Elsevier: Amsterdam, The Netherlands, 2015.

42. Lehrsch, G.A.; Kincaid, D.C. Sprinkler Irrigation Effects on Infiltration and Near-Surface Unsaturated Hydraulic Conductivity. *Trans. ASABE* **2010**, *53*, 397–404. [CrossRef]

43. Thompson, A.L.; James, L.G. Water droplet impact and its effect on infiltration. *Trans. Am. Soc. Agric. Eng.* **1985**, *28*, 1506–1510. [CrossRef]

44. Zhu, X.; Yuan, S.; Liu, J. Effect of Sprinkler Head Geometrical Parameters on Hydraulic Performance of Fluidic Sprinkler. *J. Irrig. Drain. Eng.* **2012**, *138*, 1019–1026. [CrossRef]

45. Mohammed, D.; Kohl, R.A. Infiltration response to kinetic energy. *Trans. Am. Soc. Agric. Eng.* **1987**, *30*, 108–111. [CrossRef]

agriculture

MDPI

Article

Application of Benchmarking and Principal Component Analysis in Measuring Performance of Public Irrigation Schemes in Kenya

Faith M. Muema [1],*, Patrick G. Home [2] and James M. Raude [2]

[1]   Civil Engineering Department, Pan African University Institute for Basic Sciences, Technology and Innovation (PAUSTI), P.O. Box 62000-00200 Nairobi, Kenya
[2]   Soil, Water and Environmental Engineering Department, Jomo Kenyatta University of Agriculture and Technology (JKUAT), P.O. Box 62000-00200 Nairobi, Kenya; pghome@jkuat.ac.ke (P.G.H.); ramesso@jkuat.ac.ke (J.M.R.)
*   Correspondence: faith.mawia@yahoo.com; Tel.: +254-713-009-267

Received: 29 August 2018; Accepted: 4 October 2018; Published: 12 October 2018

Abstract: The inefficient water use, and variable and low productivity in Kenyan public irrigation schemes is a major concern. It is, therefore, necessary to periodically monitor and evaluate the performance of public irrigation schemes. This prompted evaluation of performance of three rice growing irrigation schemes in western Kenya using benchmarking and principal component analysis. The aim of the study was to quantify and rank the performance of selected irrigation schemes. The performance of the irrigation schemes was evaluated for the period from 2012 to 2016 using eleven performance indicators under agricultural productivity, water supply and financial performance categories. The performance indicators were weighted using principal component analysis and combined to form a single performance score using linear aggregation method. The average performance in the Ahero, West Kano and Bunyala irrigation schemes was 48%, 49% and 56%, respectively. Based on performance score, the Bunyala irrigation scheme is the highest performing rice irrigation scheme in western Kenya. The three irrigation schemes have an average performance. Operation and management measures to improve the current performance of the irrigation schemes are needed.

Keywords: benchmarking; evaluation of performance; performance indicator; principal component analysis

1. Introduction

Irrigated agriculture occupies 4 percent of the total land area (2.9 million ha) under agriculture in Kenya [1]. It accounts for 3 percent of the Kenya's gross domestic product (GDP) and 18 percent of the total value of all agricultural produce [1]. The main irrigated crops in Kenya are rice, wheat, maize, vegetables, coffee, fruits, sugarcane, cotton and horticulture [2]. Rice is the third main cereal crop grown in Kenya after maize and wheat [3]. It is mainly grown in government-established irrigation schemes managed by National Irrigation Board (NIB). These are Ahero, Bunyala, West Kano irrigation schemes located in Western Kenya and Mwea irrigation scheme in Central Kenya. The other NIB-managed irrigation schemes are: Hola, Perkerra, Bura and, more, recently the Galana-Kulalu Food Security Project [4]. The continuous flooding method of water application is used in rice farming in Ahero, West Kano, Bunyala and Mwea. This system of rice farming utilises a lot of water, and production is highly reduced during drought periods [5]. Rice production in Kenya is below demand, and the gap is filled through imports. Currently, 54,000 metric tonnes of milled rice are produced in Kenya, whereas the current national demand for rice is 693,000 metric tonnes [6]. Rice consumption is expected to increase due to rising population, change in eating habits and urbanisation [3]. The population in Kenya has been growing rapidly, with an increase from 28.7 million in 1999 to 38.6 million in 2009, and is expected to reach 69.5 million by 2030 [4,7]. Increased demand for food and competition for water among various sectors of the economy is therefore expected. Kenya is a water-scarce country with access to 647 cubic metres of freshwater per capita per annum. This is way below the international acceptable levels of 1000 cubic metres per capita per annum [8]. Water scarcity limits water available for irrigation. Efficient utilisation of water, land and other resources increases productivity and promotes sustainable development in irrigated agriculture.

The inefficient water use, and the variable and low productivity of public irrigation schemes in Kenya is a major concern. Heavy investment is channelled into these irrigation schemes, but their productivity is below the expectation [2]. In addition, poor productivity of public irrigation schemes in Kenya hinders their expansion [9]. There is therefore a need to improve productivity and increase the efficiency of the utilisation of water and other resources. Comparative evaluation of performance using the benchmarking tool can be applied. Benchmarking is a tool used for evaluating performance of irrigation systems over time and comparing the performance with comparable irrigation systems or own set goals [10]. Comparative performance evaluation of irrigation systems enables identification of the performance gap between current and best practices [11]. The benchmarking tool was developed by the International Programme for Technology and Research in Irrigation and Drainage (IPTRID) as a management tool for improving productivity and efficiency in the irrigation and drainage sector [12]. The IPTRID, the Food and Agriculture Organisation (FAO), the World Bank, the International Water Management Institute (IWMI) and the International Commission on Irrigation and Drainage (ICID) have laid an emphasis on measuring performance in the irrigation and drainage sector as a way of achieving sustainable development in agriculture. Evaluation of the performance of irrigation schemes is based on standard performance indicators. A performance indicator is a description of actual achievement in relation to one of the goals set in an irrigation system [10]. Performance indicators can be categorised into either internal or external indicators.

External indicators examine inputs and outputs of an irrigation system [13]. The indicators describe the overall performance of irrigation systems using ratios that compare inputs to outputs. These indicators give an expression of various efficiencies related to water, budgets or yields. External indicators do not provide an insight into what should be done to improve performance. They only give an indication that improvement is needed [14]. IPTRID benchmarking indicators fall in the category of external indicators [14]. External indicators are suitable for use in cross-comparison of the performance of irrigation systems [15]. Internal indicators, on the other hand, examine the internal processes of the system and the level of water delivery service provided by the project. The indicators look into operations, hardware of the system, institutional and management set up, and water distribution and delivery [13]. Internal indicators provide an insight into what should be done to improve performance. This study was based on cross-comparison of the irrigation schemes and only external indicators were used.

Evaluation of performance of western Kenyan rice irrigation schemes was done using benchmarking indicators and principal component analysis (PCA). Comparison of performance indicators does not provide a clear picture of the overall performance of one irrigation scheme relative to others. Therefore, other tools are required when measuring the overall performance. The efficiency of nine irrigation districts in Andalusia, Spain was evaluated using performance indicators and multivariate data analysis (cluster analysis and principal component analysis) [16]. The study used principal component analysis to develop quality index for detecting performance weakness of the various irrigation districts. Also, [17] applied agglomerative hierarchical cluster analysis to group water users association (WUAs) and compared the performance of drip and sprinkler irrigation systems using performance indicators. Hierarchical cluster analysis (HCA) and data envelop analysis (DEA) was used in evaluating efficiency of performance of seventeen small and three large irrigation schemes along Senegal Valley, Mauritania [18]. The irrigation schemes were grouped into three groups using hierarchical cluster analysis and only four irrigation schemes with an average land productivity of 4.75 ton/ha were found to be technically efficient.

Principal Component Analysis (PCA)

Principal component analysis is a statistical multivariate technique that uses orthogonal transformation to convert several correlated observed variables into a smaller number of linearly uncorrelated variables known as principal components [19]. The first principal component accounts for the highest variation in data and the subsequent component has the highest variance possible, as long as it is orthogonal to the preceding component. The number of p original features is reduced into a few unobserved variables, k known as principal components. The principal components (k) account for the maximum variance such that $k \leq p$ [20]. Original features p represents the original number of observed variables for each of the case (1–n) before transformation. An example of original data with n objects and p observed variables is presented in Table 1.

Table 1. Form of data for Principal component analysis with n cases each with p features.

Case	X_1	.	X_p
1	X_{11}	.	X_{1p}
2	X_{21}	.	X_{2p}
.	.	.	.
.	.	.	.
n	X_{n1}	.	X_{np}

The principal components (Z_1, Z_2, ... , Z_i) are generated through linear combination of variables $X's$.

$$Z = \alpha^T X \tag{1}$$

where; $Z = Z_1, Z_2, Z_p$—vector of principal components; α^T-matrix of coefficients α_{ij} for $i, j = 1, 2, \ldots, p$

$$Z_1 = \alpha_{11}X_1 + \alpha_{12}X_2 + \ldots + \alpha_{1p}X_p \tag{2}$$

Z_1 is the largest combination of p features under the condition that

$$\alpha_{11}^2 + \alpha_{12}^2 + \ldots + \alpha_{1p}^2 = 1 \tag{3}$$

The second principle component Z_2 has the second-largest possible variance in $X_1, X_2, \ldots, X_p$, which is orthogonal and uncorrelated with Z_1. The jth principal component with the largest possible variance is defined similarly, provided it is uncorrelated with the ith principal component for $i < j$. The principal components obtained are in decreasing order, i.e., variance (Z_1) > variance (Z_2) > ... > variance (Z_p). If λ_i is the variance (eigenvalue) for Z_i and α_{ij} is the eigenvector for Z_i then the following conditions hold:

$$\lambda_1 \geq \lambda_2 \geq \lambda_i \geq 0 \tag{4}$$

$$\alpha_1^T \alpha_i = 1 \tag{5}$$

$$\alpha_1^T \alpha_h = 0 \tag{6}$$

The eigenvalue represents the level of variation caused by the associated principal component. The variance for the principal component for k-retained principle components is computed by

$$t_k = \frac{\sum_{i=1}^{k} \lambda_i}{\sum_{i=1}^{p} \lambda_i} \tag{7}$$

Principal components can be extracted using covariance or correlation matrix. Covariance matrix is applied where the variables do not have gross variance. For such data, standardisation of data should be done prior to using a covariance matrix. The correlation matrix, on the other hand, is applied to data with a wide variance [19]. It is suitable for analysis of variables with different measurement scales, and no prior transformation is needed. Use of the correlation matrix is not possible for data with small variance [19]. The researcher chooses the appropriate transformation matrix based on the data structure. When using the correlation matrix, only principal components with eigenvalues greater than 1 are retained. Principal components with eigenvalues greater than the average of total eigenvalues are retained when the covariance matrix is used [20]. PCA is objective and relies on the underlying data structure to generate non-subjective weights [21].

The combination of benchmarking indicators and PCA in this study enabled the description of performance using a single performance score. The performance score gives a measure of the level of performance of an individual irrigation scheme relative to the others. This study provides information to scheme managers on areas of weakness that require improvement. Furthermore, it sheds some light for stakeholders and policy makers on areas that require policy interventions.

2. Materials and Methods

2.1. Description of Study Area

The study was carried out in the Ahero, West Kano and Bunyala irrigation schemes in western Kenya managed by National Irrigation Board NIB (Figure 1). Rice is the main crop grown in these schemes. In all the schemes, water is abstracted using electric-powered pumps, conveyed with open earth canals and applied using basin irrigation method. Drain water is pumped back to Lake Victoria in the West Kano irrigation scheme because the outlet is on lower ground than the lake. The schemes have no gauging stations. Western Kenya is hot and humid, with a bimodal rainfall pattern. The schemes are underlain by deep black cotton soils [22].

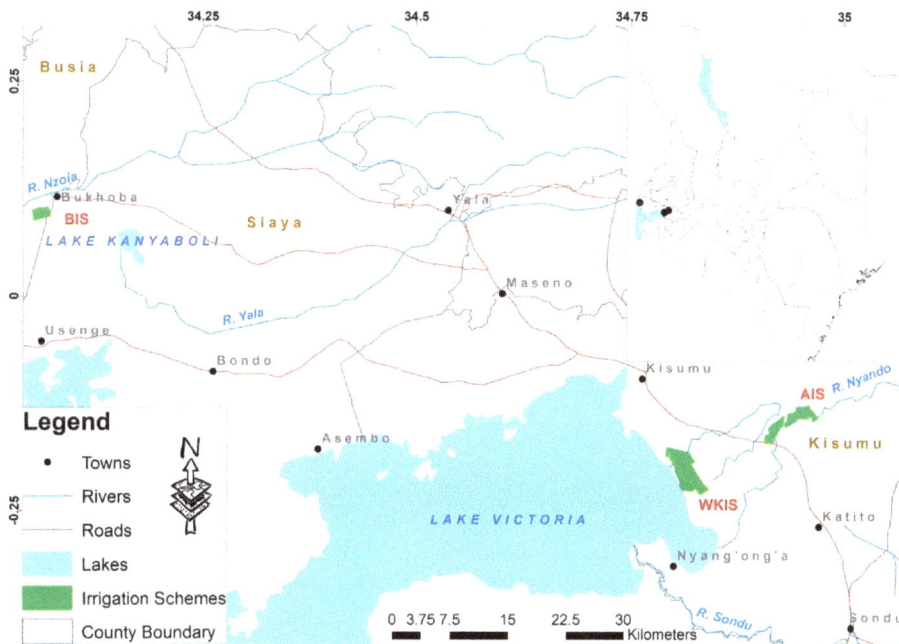

Figure 1. Study Area.

A detailed description of the main features of three irrigation schemes studied is presented in Table 2.

Table 2. Main characteristics of irrigation schemes benchmarked in western Kenya.

Description	Irrigation Scheme		
	Ahero	West Kano	Bunyala
Command area (ha)	900	980	728
Latitude	00°10′ South	Between 00°04′ South and 00°20′ South	00°06′ North
Longitude	34°58′ East	Between 34°48′ East and 35°02′ East	34°04′ East
Location	Kano plains, Kisumu county	Kano plains, Kisumu county	Kisumu/Siaya county
Land ownership	Government	Government	Government and private
Main crops	Rice, Soybeans, maize, Watermelon, sorghum	Rice, sorghum, maize	Rice, pulses and horticulture
Number of seasons	2 seasons 1st season-rice 2nd-other crop	2 seasons 1st season-rice 2nd-other crop	2 seasons 1st season-rice 2nd-other crop
Number of farmers	556	845	1934
Farm size (acres)	1–4	2–4	1–5
Water source	Surface water River Nyando	Surface water Lake Victoria	Surface water River Nzoia
Type of water distribution	On demand	On demand	On demand
Method of water abstraction	Pumping using electricity 2 pumps each 1100 L/s 2 pumps each 650 L/s	Pumping using electricity 3 pumps each 750 L/s	Pumping using electricity 4 pumps each 300 L/s
Water delivery infrastructure	Open earth canals	Open earth canals	Open earth canals
Type of water control equipment	None	None	None
Discharge measurement facilities	None	None	None
Irrigation system	Surface-Basin	Surface-Basin	Surface-Basin
Water availability	Sufficient-occasionally not sufficient	Abundant	Abundant
Type of surface drain	Open earth channel by gravity	Pumped through open earth channel. Using four 500 L/s outlet pumps.	Open earth channel
Type of revenue collection	Charge on irrigated area	Charge on irrigated area	Charge on irrigated area

2.2. Data Collection

Secondary time series data for five years (2012–2016) was obtained from records kept by management of the various irrigation schemes. The data collected was only for rice production. Rice is grown in the first season, while the other crops are grown in the second season. The production of the other crops has not been formalised, and their production is not documented. Data on total yield per season, local crop price per season, cropped area, total command area, revenue collected, expected revenue, cost of production, water supplied, pump speed, and pumping hours was collected from records kept by the irrigation scheme offices and field survey. Meteorological data was obtained from Ahero research station, West Kano weather station, the Kenya Meteorological Department (KMD) and the NASA POWER Centre. Key informant interviews, observation, and focus group discussion methods were used to collect data on farming practices, cropping pattern, status of the irrigation systems and maintenance of the system.

2.3. Data Analysis

Field data was first processed to obtain variables for calculating performance indicators. The variables were computed as follows:

(a)   Crop water requirement

Crop pattern, transplanting date and weather data was used in calculating rice crop water demand and crop irrigation water requirement using CROPWAT 8.0 software (developed by FAO, Rome, Italy). Computation of reference crop water demand (ET_0) is based on the Penman Monteith equation. The effective rainfall was computed using USDA-Soil Conservation Method, in-built in CROPWAT 8. Number of sunshine hours, temperature, humidity, rainfall data, wind speed, soil type, transplanting date and crop pattern were used as input for the model. The total annual volume of water consumed by all crops in the irrigation schemes was computed using Equation (8) [10].

$$VEt_c = \sum\nolimits_{crops} Et_c \times A \tag{8}$$

VEt_c = Total volume of crop water demand (m^3); Et_c = crop evapotranspiration from planting to harvesting (m^3); A = cropped area.

Total annual volume crop irrigation demand was then calculated using Equation (9) [10].

$$VEt_{Net} = IR_n A \tag{9}$$

VEt_{Net} = Total volume of water consumed by crops less effective rainfall (m^3); IR_n = net irrigation water requirement (m^3); A = cropped area

(b)   Total annual volume of irrigation water supply (m^3). This was obtained by summing the daily volume of water pumped for the rice growing season in each year. Daily volume of water pumped was obtained as the product of pump efficiency, pumping hours and the pump operating speed.

(c)   Total annual volume of water supply (m^3). This was obtained by summing the total volume of water pumped for irrigation and total effective rainfall for the rice growing season in a year. The effective rainfall was computed using the USDA-Soil Conservation Method, in-built in CROPWAT 8. The effective rainfall in terms of depth was converted into volume by multiplying by the total annual cropped area.

(d)   Total annual cropped area (ha). This was calculated by summing up all the area under rice crop in each year.

(e)   Total command area of the system (ha). This is the net area serviced by the scheme less the right of way for canals, drains, roads and villages. It was obtained from the design office of each irrigation scheme.

The performance indicators used were obtained from the IPTRID benchmarking indicators presented in Table 3 [10].

Table 3. Proposed key performance indicators.

Domain	Performance Indicator	Data Required
	Total annual volume of irrigation water delivery (m^3/year)	Total daily measured water delivery to water users
Service delivery performance	Annual irrigation water delivery per unit command area (m^3/ha)	Total daily measured water inflow to the irrigation system
		Total command area serviced by the system
	Annual irrigation water delivery per unit irrigated area (m^3/ha)	Total daily measured water inflow to the irrigation system
		Total annual irrigated crop area

Table 3. *Cont.*

Domain	Performance Indicator	Data Required
	Main system water delivery efficiency	Total daily measured water delivery to water users
		Total daily measured water inflow to the irrigation system
	Annual relative water supply	Total daily measured water inflow to the irrigation system
		Total daily measured rainfall over irrigated area
		Total daily/periodic volume of crop water demand, including percolation losses for rice crops
	Annual relative irrigation supply	Total daily measured water inflow to the irrigation system
		Total daily/periodic volume of irrigation water demand (crop water demand excluding effective rainfall), including percolation losses for rice
	Water delivery capacity	Current main canal capacity
		Peak month irrigation water demand
	Security of entitlement supply	System water entitlement
		10 years minimum water availability flow pattern
Financial performance	Cost recovery ratio	Total revenues collected from water users
		Total management, operation and maintenance (MOM) cost
	Maintenance cost to revenue ratio	Total maintenance expenditure
		Total revenue collected from water users
	Total MOM cost per unit area (US$/ha)	Total management, operation and maintenance expenditure
		Total command area serviced by the system
	Total cost per person employed on water delivery (US$/person)	Total cost of MOM personnel
		Total number of MOM personnel employed
	Revenue collection performance	Total revenues collected from water users
		Total service revenue due
	Staffing numbers per unit area (persons/ha)	Total number of MOM personnel employed
		Total command area serviced by system
	Average revenue per cubic meter of irrigation water supplied (US$/m^3)	Total revenues collected from water users
		Total daily measured water delivery to water users
Agricultural Productive efficiency	Total gross annual agricultural production (tones)	Total tonnage produced under each crop
	Total annual value of agricultural production (US$)	Total annual tonnage of each crop
	Output per unit serviced area (US$/ha)	Crop market price
		Total annual tonnage of each crop
		Crop market price
		Total command area serviced by system
	Output per unit irrigated area (US$/ha)	Total annual tonnage of each crop
		Crop market price
		Total annual irrigated crop area
	Output per unit irrigation supply (US$/m^3)	Total annual tonnage of each crop
		Crop market price
		Total daily measured water inflow to the irrigation system

Table 3. *Cont.*

Domain	Performance Indicator	Data Required
	Output per unit water consumed (US$/m^3)	Total annual tonnage of each crop
		Crop market price
		Total volume of water consumed by the crops (ET_c)
	Water quality: Salinity (mmhos/cm)	Total daily measured water inflow to the irrigation system
		Electrical conductivity of periodically collected drainage water samples
		Total daily measured drainage water outflow from the irrigation system
	Water quality: Biological (mg/litre)	Biological load of periodically collected irrigation water samples
		Total daily measured water inflow to the irrigation system
		Biological load of periodically collected drainage water samples
Environmental performance		Total daily measured drainage water outflow from the irrigation system
	Water quality: Chemical (mg/litre)	Chemical load of periodically collected irrigation water samples
		Total daily measured water inflow to the irrigation system
		Chemical load of periodically collected drainage water samples
		Total daily measured drainage water outflow from the irrigation system
	Average depth to water table (m)	Periodic depth measurement to water table
	Change in water table depth over time (m)	Periodic depth measurement to water table over 5 year period
	Salt balance (tones)	Periodic measurement of salt content of irrigation water
		Periodic measurement of salt content of drainage water

The methodology adopted entails: (i) selection of suitable indicators to describe performance of the irrigation schemes; (ii) combining the indicators into a single performance score using principal component analysis. Some of the proposed key performance indicators (Table 3) were not computed because of lack of data.

2.3.1. Performance Indicators

Fourteen performance indicators were computed as shown in Table 4. The indicator values were compared among the three schemes in each year.

To allow for global comparison, the total value of agricultural production is converted into gross value of production using Equation (10).

$$GVP = \left\lfloor \sum crops\, A_i Y_i \right\rfloor \text{ MU currency exchange rate} \qquad (10)$$

GVP—gross value of production; A_i—area cropped with crop *i*; Y_i—the yield of crop *i*; P_i—local price of crop *i*; *MU*—currency exchange rate (US$ per unit local currency).

Table 4. Computation of performance indicators.

Performance Indicator	Definition/Calculation
Total annual volume irrigation supply	Total annual volume of irrigation water pumped or diverted
Annual relative water supply	$\dfrac{\text{total annual volume of water supply}}{\text{total annual volume of crop water demand}}$
Annual relative irrigation supply	$\dfrac{\text{total annual volume of irrigation supply}}{\text{total annual volume of crop irrigatio demand}}$
Annual irrigation supply per unit irrigated area (m^3/ha)	$\dfrac{\text{Total annual volume of irrigation supply}}{\text{Total annual irrigated area}}$
Annual irrigation supply per unit command area (m^3/ha)	$\dfrac{\text{Total annual volume of irrigation supply}}{\text{Total annual command area}}$
Total gross annual agricultural production (tones)	Total annual tonnage of each crop
Total annual valueof agricultural production (US$)	Total annual gross value of production (GVP) received by producers
Output per unit irrigated area (US$/ha)	$\dfrac{\text{Total annual value of agricultural production}}{\text{Total annual irrigated area}}$
Output per unit command area (US$/ha)	$\dfrac{\text{Total annual value of agricultural production}}{\text{Total command area}}$
Output per unit water supply (US$/m^3)	$\dfrac{\text{Total annual value of agricultural production}}{\text{total annual volume of water suply}}$
Output per unit irrigation supply (US$/m^3)	$\dfrac{\text{Total annual value of agricultural production}}{\text{total annual volume of irrigation suply}}$
Output per unit crop water demand (US$/m^3)	$\dfrac{\text{Total annual value of agricultural production}}{\text{total annual volume of crop water demand}}$
Water fee collection performance (%)	$\dfrac{\text{Gross revenue collected}}{\text{Gross revenue invoiced}} \times 100$
Average revenue per unit irrigation supply (US$/m^3)	$\dfrac{\text{total annual revenue collected}}{\text{Total annual volume of irrigation supply}}$

2.3.2. Calculation of Overall Irrigation Scheme Performance

The overall scheme performance was determined by computing a single performance score. The total volume of irrigation water supply, total annual agricultural production and total annual value of agricultural production indicators were excluded in the computation of overall performance score. These indicators are based on extensive scale rather than relative scale and their inclusion might distort the results. Indicators were first tested for statistical correlation using the Pearson correlation method. Ten indicators with low correlation were selected. The indicators were weighted using principal component analysis, then normalised using the reference to target method and finally aggregated into a single performance score using the linear aggregation method. Weighting of indicators was done using PCA.

Principal Component Analysis (PCA)

PCA was done using SPSS windows version16 software. Prior to PCA, the data was tested for suitability using Kaiser-Meyer-Olklin (KMO) and Bartlett Test of Sphericity (BTS). The extracted components were rotated using orthogonal varimax method to achieve significant components. The indicator weights were computed using rotated factor loadings and eigenvalues, as shown in Equation (11).

$$W_k = \sum_{j=1}^{j=n} \frac{(\text{Factor loading}_{kj})^2}{\text{eigenvalue}_j} \times \frac{\text{eigenvalue}_j}{\sum_{j=1}^{j=n} \text{eigenvalue}_j} \tag{11}$$

Factor loading$_{kj}$—factor loading of indicator k in the principal component j; eigenvalue$_j$—eigenvalue for j^{th} principal component; $j = 1, j = 2, \dots, j = n$ the extracted principal components with an eigenvalue above 1.

The indicators were normalised using reference to target using Equation (12).

$$I_{qs}^t = \frac{x_{qs}^t}{x_b} \tag{12}$$

I_{qs}^t = normalised value of indicator q for scheme s at time t; x_{qs}^t = indicator value for scheme s at time t; x_b = threshold value for indicator value.

The threshold values used for normalisation of indicators are shown in Table 5.

Table 5. Indicative threshold values.

Performance Indicator	Threshold Values	Reference
Relative water supply	2	[23]
Relative irrigation supply	2	[23]
Annual irrigation water delivery per unit irrigated area	450–700 mm	[24]
Annual irrigation water delivery per unit command area	450–700 mm	[24]
Output per unit irrigated area	3.8 ton/ha	[25]
Output per unit command area	3.8 ton/ha	[25]
Output per unit irrigation supply	2 kg/m³	[26]
Output per unit water supply	2 kg/m³	[26]
Output per water consumed	2 kg/m³	[26]
Water fee collection performance	100%	[10]
Average revenue per unit irrigation supply	7.5 US dollar cents	[27]

A single performance score was finally computed using Equation (13).

$$CI_{st} = \sum_{k=1}^{k=n} W_k I_{ks} \tag{13}$$

where; W_k = indicator weight; I_{ks} = normalised indicator k for scheme s; CI_{st} = performance score for irrigation scheme s at time t.

3. Results and Discussion

The results of comparative evaluation of performance using performance indicators are presented as follows.

3.1. Water Supply Performance

The indicators under this category give a measure of water supply relative to demand. Water abundance or scarcity of water can be deduced from these indicators [28]. The results of water supply indicators are presented in Table 6. The command area and irrigated area used in computation of various performance indicators for each scheme is also presented in Table 6.

The available irrigable area (command area) in all the schemes has not been fully exploited. Some of the command area is not irrigated due to the inability of farmers to acquire farming inputs. Irrigated area in Ahero and Bunyala irrigation schemes is close to command area. The low irrigated area in West Kano in 2013 and 2014 can be attributed to lack of interest in irrigation by farmers following the collapse of the revolving fund committee. The annual volume of irrigation supply for the schemes ranges between 2.2 and 8.4 MCM. All the schemes divert water by pumping using electricity. The amount of water abstracted at any given time depends on cropped area. The irrigation schemes have a high fluctuation in the amount of water supplied due to frequent power outages experienced in the region. The amount of water abstracted is estimated using pumping hours recorded, pump speed and pumping efficiency. The amount of water delivered to irrigation blocks could not be computed.

Table 6. Water supply indicators.

Irrigation Scheme	Year	Command Area (ha)	Total Annual Irrigated Area (ha)	Total Annual Volume of Irrigation Water Supply (m³)	RWS	RIS	Annual Water Deliver per Unit Irrigated Area (m³/ha)	Annual Water Delivery per Unit Command Area (m³/ha)
Ahero	2012/2013	900	877	6,827,820	1.98	2.15	7785	7586
	2013/2014	900	846	4,938,460	1.14	0.86	5837	5487
	2014/2015	900	783	4,867,840	1.45	1.31	6217	5409
	2015/2016	900	824	4,362,330	1.28	0.86	5294	4847
	2016/2017	900	720	3,950,460	1.24	0.68	5487	4389
Average			810	4,989,382	1.42	1.17	6124	5544
West Kano	2012/2013	980	617	6,934,097	2.31	3.38	11,238	7076
	2013/2014	980	206	2,540,691	1.94	1.64	12,310	2593
	2014/2015	980	196	2,223,590	2.21	2.74	11,376	2269
	2015/2016	980	650	7,115,034	1.92	1.75	10,955	7260
	2016/2017	980	690	8,411,680	1.86	1.58	12,191	8583
Average			472	5,445,018	2.05	2.22	11,614	5556
Bunyala	2012/2013	728	701	4,406,847	1.98	1.94	6287	6050
	2013/2014	728	701	6,215,776	2.17	2.25	8868	8533
	2014/2015	728	701	5,401,296	2.06	2.26	7706	7415
	2015/2016	728	625	5,387,886	2.24	2.40	8622	7396
	2016/2017	728	666	8,077,590	2.44	2.46	12,130	11,089
Average			679	5,897,879	2.18	2.26	8723	8097

RWS—Relative Water Supply; RIS—Irrigation Water Supply.

The relative irrigation supply (RIS) values varied from 0.68 to 3.38 during the study period. RIS and RWS values should be above 1. This is because irrigation efficiency is always below 100% due to unavoidable conveyance and application losses. Values below 1 indicate water deficit [26,27]. The average RIS in the Ahero, west Kano and Bunyala irrigation schemes was 1.17, 2.22 and 2.26, respectively. A low RIS value of 0.4 was reported in Muda irrigation scheme, Malaysia [15]. The low RIS was associated with the use of real-time monitoring of water depth in rice farms, which enabled effective use of rainfall. The relative water supply (RWS) varied between 1.14 and 2.44 for all the schemes. RWS above 2 shows that the amount of water supplied is adequate [15]. High RIS and RWS values in the West Kano and Bunyala irrigation schemes show that there is adequate supply of water. The Ahero irrigation scheme suffers from inadequate supply of water, which is evident from the low RIS values, the majority of which are below 1. The Ahero irrigation scheme draws water from the river Nyando, which is occasionally affected by drought and siltation. The Ahero irrigation scheme was in drought, which lowered the amount of water available for irrigation in 2013. In 2016, one of the water pumps, with a discharge capacity (100 L/s), broke down. This contributed to a very low RIS of 0.68. The water shortage in all the irrigation schemes is due to frequent power outages.

The average RIS values obtained are comparable to the average RIS value of 2.31 recorded in the large public rice irrigation schemes in the Senegal Valley in Mauritania [18]. An average RWS of 0.77 was obtained in Karacabey surface irrigation system, Turkey [29]. This irrigation scheme was reported to have a water shortage. Elsewhere in Turkey, [30] obtained RWS values ranging between 0.37 and 1.97. In Malaysia, RWS varied between 0.4 and 4.3, while RIS ranged between 0.5 and 5.7. The high values in Malaysia are attributed to extensive rice farming using open channels. An average RIS value of 1.38 was obtained for sprinkler irrigation systems and 1.03 for drip irrigation systems in Spain [17]. Sprinkler and drip irrigation systems have a high irrigation efficiency compared to the surface irrigation method. That is why the RIS values are lower compared to the values obtained in the Ahero, Bunyala and West Kano irrigation schemes.

The quantity of water supplied per unit area varies with the availability of water, climate, soil type, cropping pattern, system conditions and system management [31]. The annual water delivery per unit command area (WDCA) varied between 2269 m^3/ha (West Kano in 2014/2015) to 11,089 m^3/ha (Bunyala in 2016/2017). The WDCA was 4389 m^3/ha−7586 m^3/ha in Ahero, 2269 m^3/ha−8583 m^3/ha in West Kano and 6050 m^3/ha−11,089 m^3/ha in the Bunyala irrigation scheme. WDCA was highest in Bunyala, and least in the Ahero irrigation scheme. The annual water delivery per unit irrigated area (WDIA) varied from 5294 m^3/ha to 7785 m^3/ha in Ahero; 11,238 m^3/ha to 12,310 m^3/ha in West Kano and 6285 m^3/ha to 12,130 m^3/ha in the Bunyala irrigation scheme. This is equivalent to supplied depth of water of 529.4 mm−778.5 mm in Ahero, 1123.8 mm−1231 mm in West Kano, and 628.5 mm to 12,130 mm in the Bunyala irrigation scheme. According to the FAO, the average crop water needed for paddy rice should be 450 mm−700 mm [32]. Considering low irrigation efficiencies associated with surface irrigation schemes—usually 30–40% [32]—the WDIA is adequate in the West Kano and Bunyala irrigation schemes. The Ahero irrigation scheme, on the other hand, supplies inadequate water, which is not enough to meet crop water needs. WDIA values are relatively lower compared to the 22,029.43 m^3/ha, 16,026.37 m^3/ha, 11,289.10 m^3/ha, and 9795.96 m^3/ha obtained in MARIIS, Divisoria, Lucban and Garab SWIPs, respectively, in the Cagayan river basin, Philippines [23]. In southern Italy, high WDIA values ranging between 6500–14,900 m^3/ha were reported by the Water Users' Association (WUA's) of Calabria [33]. WDIA values of 5578 m^3/ha were obtained in sprinkler irrigation systems and 1084 m^3/ha in drip irrigation systems in Castilla-La Mancha, Spain [17]. These values are much lower than the values obtained in this study. Drip and sprinkler irrigation systems have high water application efficiencies of 75% and 90%, respectively [34]. Surface irrigation systems, on the other hand, have a low irrigation efficiency of 60%. Therefore, more water is supplied in surface irrigation systems compared to sprinkler and drip irrigation systems. In the Susurluk river basin in Turkey, WDCA values varying from 1465 m^3/ha to 13,086 m^3/ha and WDCA values ranging from 2169 m^3/ha to 22,098 m^3/ha were obtained [30]. A high amount of water is supplied to irrigation schemes in the Sursurluk basin because rainfall is limited during the irrigation period.

3.2. Financial Performance

The financial performance indicators measure the efficiency with which irrigation systems use resources to provide service to farmers [23]. The results are shown in Table 7.

Table 7. Financial performance indicators.

		Gross Revenue Collected (US$)	Gross Revenue Invoiced (US$)	Revenue Collection Performance (%)	Average Revenue per Unit Irrigation Water Supply (US Dollar Cents /m^3)
	2012/2013	24,979.50	55,510.00	45	0.36
	2013/2014	8910.72	18,564.00	48	0.35
WKIS	2014/2015	8966.41	17,581.20	51	0.40
	2015/2016	31,547.88	58,422.00	54	0.44
	2016/2017	35,375.34	62,062.00	57	0.42
	Average	21,955.97	42,427.84	51	0.39
	2012/2013	67,208.00	53,766.40	80	0.79
	2013/2014	64,790.00	55,071.50	85	1.12
AIS	2014/2015	59,985.00	49,187.70	82	1.01
	2015/2016	63,147.00	54,306.42	86	1.24
	2016/2017	55,180.00	49,662.00	90	1.26
	Average	62,062.00	52,398.80	85	1.08
	2012/2013	69,280.00	63,737.60	92	1.45
	2013/2014	69,280.00	65,123.20	94	1.05
BIS	2014/2015	69,280.00	64,430.40	93	1.19
	2015/2016	61,770.00	58,681.50	95	1.09
	2016/2017	65,820.00	63,811.20	97	0.79
	Average	67,086.00	63,156.78	94	1.11

Currency exchange rate—1 US$ = 100 Kenya shilling (KES).

Water fee collection performance (WFC) values obtained are 80–90% in the Ahero irrigation scheme, 45–57% in the West Kano irrigation scheme, and 92–97% in the Bunyala irrigation scheme. According to [18], water fee collection values below 70% are considered unsatisfactory. Bunyala has the highest average fee collection performance, at 94%, while West Kano has the lowest average value, at 51%. The ideal desirable value should be close to 100% [10]. De Alwis and Wijesekara [35] obtained an ideal WFC of 100% in the Beypazarı Başören irrigation system, Turkey. Similarly, a WFC of 103% was recorded in the Karacabey irrigation scheme in Turkey. Values of WFC equal to or above 100% show that water users are willing to pay for the cost of irrigation. WFC values above 100% are possible to obtain due to payment of accumulated arrears. Low WFC values point out an unwillingness of farmers to pay water fees, poor organisation of the Irrigation Water Users Association (IWUA), poor collection programs, and financial problems within the schemes. Bunyala is able to sustain a value above 90% because of the well-organised farmer groups that are mandated with the mobilisation of the water fee. Also, in Bunyala, the policy of water fee payment prior to ploughing is strictly followed.

The average revenue per unit cubic meter varied from 0.79 to 1.26 US cents in the Ahero irrigation scheme, 0.35 to 0.44 US cents in the West Kano irrigation scheme, and 0.79 to 1.45 US cents in the Bunyala irrigation scheme. These values are below the economic value of irrigation water of 7.54 US cents per cubic meter obtained by [27] in the Ahero irrigation scheme. Pricing of water is an economic aid to improving water allocation and sustainable water utilisation [30]. The water fee charged is US$31, US$36.40 and US$40 per acre in the Ahero, West Kano and Bunyala irrigation schemes, respectively. The pricing is based on area cropped per farming season and not the quantity of water consumed. There is no limit to the quantity of water that a farmer can use. This explains why the value of water per cubic meter is below 1 US$. This is a weakness and is unsuitable in terms of efficiency of water use and water conservation. Bunyala is the best performing irrigation scheme under the financial performance category.

3.3. Agricultural Productivity

Agricultural productivity gives the relationship between inputs and output. It gives an indication of efficiency of crop production in terms of land used, amount of water used and the income generated [36]. The indicators are presented in Table 8.

Table 8. Agricultural productivity indicators.

Irrigation Scheme	Year	AGP (Tones)	GVP (US$)	OIA (US$/ha)	OCA (US$/ha)	OIS (US$/m^3)	OWS (US$/m^3)	OCWD (US$/m^3)
West Kano	2012/2013	2679	857,120	950	1389	0.12	0.09	0.21
	2013/2014	1201	420,308	466	2036	0.17	0.13	0.26
	2014/2015	1136	374,959	416	1918	0.17	0.12	0.27
	2015/2016	3633	1,307,837	1450	2014	0.18	0.14	0.26
	2016/2017	4083	1,551,707	1720	2249	0.18	0.15	0.28
Average		2546	902,386	1000	1921	0.17	0.13	0.26
Ahero	2012/2013	4179	1,677,200	1864	1912	0.25	0.14	0.28
	2013/2014	4182	1,479,800	1644	1749	0.30	0.20	0.22
	2014/2015	4551	1,683,870	1871	2151	0.35	0.20	0.29
	2015/2016	4465	1,741,370	1935	2113	0.40	0.21	0.27
	2016/2017	4058	1,663,780	1849	2311	0.42	0.22	0.28
Average		4287	1,649,204	1832	2047	0.34	0.20	0.27
Bunyala	2012/2013	3803	1,248,221	1714	1781	0.28	0.15	0.29
	2013/2014	2146	714,678	981	1020	0.11	0.07	0.16
	2014/2015	3380	1,132,300	1554	1615	0.21	0.13	0.26
	2015/2016	3850	1,321,617	1814	2115	0.25	0.15	0.34
	2016/2017	3633	1,214,772	1668	1824	0.15	0.11	0.28
Average		3362	1,126,318	1546	1671	0.20	0.12	0.27

AGP—annual gross agricultural production; GVP—gross value of agricultural production; OIA—output per unit irrigated area; OCA—output per unit command area; OIS—output per unit irrigation supply; OCWD—output per unit crop water demand; Currency exchange rate—1 US$ = 100 Kenya shilling (KES).

The output per unit irrigation supply (OIS) ranges between 0.11 US$/m^3 and 0.42 US$/m^3 in all the schemes. The average, OIS is 0.34 US$/m^3 in Ahero, 0.17 US$/m^3 in West Kano, and 0.20 US$/m^3 in the Bunyala irrigation scheme. Ahero irrigation scheme utilises water more efficiently compared to the others. The output per unit water supply (OWS) puts into consideration the contribution of effective rainfall. The values vary between 0.07 and 0.22 US$/m^3. The highest OCWD value (0.34 US$/m^3) was obtained in the Bunyala irrigation scheme in 2015/2016, while the lowest value (0.16 US$/m^3) was recorded in Bunyala in 2013/2014. The Ahero and Bunyala irrigation schemes have the highest average OCWD of 0.27 US$/m^3, while West Kano irrigation scheme has the lowest average value of 0.26 US$/m^3. The Ahero irrigation scheme is leading in terms of water productivity while West Kano is the poorest. According to [15], if OCWD is greater than OIS, some of the irrigation water supplied is unproductive. In both West Kano and Bunyala, OIS is greater than OCWD. This shows inefficient use of water. The Ahero irrigation scheme is the most efficient water user, with all OIS values less than OCWD except in 2012/2013. The difference in water productivity is brought about by differences in yield and crop market price. In similar studies in Malaysia, [37] reported OWS values ranging between 0.01 US$/m^3 and 0.2 US$/m^3 and OCWD values varying from 0.01 US$/m^3 to 0.4 US$/m^3 for paddy rice. Compared to this, the Ahero, West Kano and Bunyala irrigation schemes registered higher rice water productivity. The difference is attributed to the yield and market price. Mchele [38] obtained OIS values of 0.95 US$/m^3 in Shina-Hamusit and 0.62 US$/m^3 in the Selamko irrigation scheme, and OCWD values of 1.46 US$/m^3 in Shina-Hamusit and 1.15 US$/m^3 in the Selamko irrigation scheme, Ethiopia. These irrigation schemes do not grow rice. This shows that rice is a competitive crop in terms of returns per water used. In Turkey, OCWD values varying between 0.191 US$/m^3 and 1.262 US$/m^3

were obtained [30]. The highest values of rice water productivity of 1.77 kg/m³, 1.75 kg/m³ and 1.51 kg/m³ have been reported in the USA, Sri Lanka and Spain, respectively [26].

Land productivity indicators give a reflection of crop intensity [39]. The output per unit command area (OCA) varies between 1020 US$/ha (Bunyala in 2013/2014) and 2311 US$/ha (Ahero in 2016/2017). The average OCA computed was 2047 US$/ha in Ahero; 1921 US$/ha in West Kano and 1671 US$/ha in the Bunyala irrigation scheme. A high value is an indication of intensive irrigation. The sudden fall in output per command area in West Kano between 2012 and 2014 can be attributed to the collapse of the Revolving Fund Committee. The committee was mandated with the responsibility for production and marketing in the West Kano irrigation scheme. Consequently, there was a decline in production activities during that period associated with governance issues. From 2015 each block in the scheme established a production management structure which induced competition amongst the blocks in terms of production activities. An increase in production was therefore realised in 2015/2016. The Bunyala irrigation scheme experienced hail in 2013 which shattered mature rice crops in one of the phases (Muluwa phase 1). This contributed to a low harvest, as depicted by the sudden decline in the output per unit area in the scheme. The output per unit irrigated area (OIA) for all the schemes varied from 981 US$/ha to 1841 US$/ha. The OIA values computed are comparable to the OIA values of 1300 US$/ha and 1310 US$/ha obtained during the rainy season and dry season in rice farming in Thailand [40]. OIA values ranging from 100 US$/ha to 800 US$/ha were reported in Malaysia [37].

3.4. Estimation of Overall Scheme Performance

Correlation analysis of the 11 selected indicators is presented in Table 9. RWS and RIS are strongly positively correlated ($r = 0.950$). This means that the indicators measure similar elements. To avoid double counting, only one of them can be used in the computation of the composite indicator/performance score. RIS focuses on irrigation water supply alone and is therefore used for computation of the performance score.

Table 9. Pearson correlation matrix (n).

Variables	RWS	RIS	WDIA	WDCA	WFC	RIWS	OIA	OCA	OIS	OCWD	OWS
RWS	1	0.950	0.711	0.458	−0.168	−0.425	−0.185	−0.305	−0.801	0.278	−0.855
RIS	0.950	1	0.648	0.455	−0.259	−0.457	−0.273	−0.292	−0.778	0.189	−0.845
ISIA	0.711	0.648	1	0.283	−0.645	−0.873	0.092	−0.570	−0.870	0.060	−0.701
ISCA	0.458	0.455	0.283	1	0.284	−0.029	−0.312	0.475	−0.432	−0.047	−0.471
WFC	−0.168	−0.259	−0.645	0.284	1	0.889	−0.180	0.696	0.469	0.125	0.278
ARIS	−0.425	−0.457	−0.873	−0.029	0.889	1	−0.125	0.662	0.723	0.109	0.514
OIA	−0.185	−0.273	0.092	−0.312	−0.180	−0.125	1	0.156	0.375	0.741	0.553
OCA	−0.305	−0.292	−0.570	0.475	0.696	0.662	0.156	1	0.566	0.301	0.480
OIS	−0.801	−0.778	−0.870	−0.432	0.469	0.723	0.375	0.566	1	0.217	0.936
OWC	0.278	0.189	0.060	−0.047	0.125	0.109	0.741	0.301	0.217	1	0.245
OWS	−0.855	−0.845	−0.701	−0.471	0.278	0.514	0.553	0.480	0.936	0.245	1

RWS—relative water supply; RIS—relative irrigation supply; WDIA—Water delivery per unit irrigated area; WDCA—water delivery per unit command area; WFC—water fee collection; RIWS—annual revenue per unit irrigation water supply; OIA—output per unit irrigated area; OCA—output per unit command area; OIS—output per unit irrigation supply; OCWD—output per unit crop water demand; OWS—output per unit water supply.

Principal Component Analysis

The extracted principal factors, Kaiser-Meyer-Olkin (KMO) measure of sampling adequacy and Bartlett's sphericity test (BTS) results are presented in Table 10.

According to [41], if the KMO value is greater than 0.5 and the BTS less than 0.05, the data is suitable for PCA. In this study, the KMO co-efficient of 0.510 is adequate and the Bartlett's test is significant at 99% ($p < 0.0001$). The principal components extracted with their factor loadings are presented in Table 11. The indicator weights are also presented in this Table 11.

Table 10. Kaiser-Meyer-Olkin (KMO) and Bartlett's test.

KMO and Bartlett's Test		
Kaiser-Meyer-Olkin Measure of Sampling Adequacy.		0.510
Bartlett's Test of Sphericity	Approx. Chi-Square	211.443
	df	45
	Sig.	0.000

Table 11. The extracted principal components.

Rotated Component Matrix (Factor Loading)				Indicator Weights
	Principal Component			
	1	2	3	
% of variance	34.959	34.918	19.930	
Relative irrigation supply (RIS)	−0.222	−0.876	0.044	0.091
Water delivery per unit irrigated area (WDIA)	−0.673	−0.669	0.169	0.104
Water delivery per unit command area (WDCA)	0.447	−0.794	−0.047	0.093
Water fee collection performance (WFC)	0.924	0.064	−0.068	0.096
Average revenue per unit irrigation water supply (RIWS)	0.862	0.387	−0.086	0.100
Output per unit irrigated area (OIA)	−0.161	0.317	0.912	0.107
Output per unit command area (OCA)	0.891	0.040	0.289	0.098
Output per unit irrigation supply (OIS)	0.505	0.815	0.229	0.108
Output per crop water demand (OCWD)	0.149	−0.081	0.925	0.098
output per unit water supply (OWS)	0.318	0.848	0.353	0.105

Extraction Method: Principal Component Analysis.
Rotation Method: Varimax with Kaiser Normalisation.

The first principal component (PC1) determines 34.9595% of the total variance in performance. The first principal component is mainly linked to indicators with absolute factor loading greater than 0.673 (WDIA, WFC, RIWS, OCA). The second principal component (PC2) accounts for 34.918% of the variance in performance. It is influenced by RIS, WDCA, OIS and OWS indicators (absolute loadings > 0.794). The third principal component (PC3) factor loading accounts for 19.93% and is linked with OIA and OCWD indicators.

The results of weighted indicators are presented in Table 12. The performance score for each scheme in each year was obtained by summing up the weighted indicator values.

Table 12. Weighted performance score for each category.

Year	IS	RIS	WSIA	WSCA	WFC	ARIWS	OIA	OCA	OIS	OCWD	OWS	PS
	AIS	0.10	0.07	0.06	0.08	0.01	0.06	0.05	0.03	0.03	0.02	0.52
2012/2013	WKIS	0.14	0.10	0.06	0.04	0.00	0.05	0.03	0.02	0.03	0.02	0.51
	BIS	0.09	0.06	0.05	0.09	0.02	0.07	0.06	0.05	0.04	0.02	0.54
	AIS	0.04	0.05	0.05	0.08	0.01	0.06	0.05	0.05	0.03	0.03	0.45
2013/2014	WKIS	0.07	0.11	0.02	0.05	0.00	0.07	0.01	0.03	0.04	0.02	0.43
	BIS	0.10	0.08	0.07	0.09	0.01	0.04	0.03	0.02	0.02	0.01	0.48
	AIS	0.06	0.06	0.04	0.08	0.01	0.07	0.06	0.05	0.04	0.03	0.49
2014/2015	WKIS	0.11	0.10	0.02	0.05	0.01	0.07	0.01	0.03	0.04	0.02	0.46
	BIS	0.10	0.07	0.06	0.09	0.02	0.06	0.05	0.03	0.04	0.02	0.54
	AIS	0.04	0.05	0.04	0.08	0.02	0.07	0.06	0.06	0.03	0.03	0.47
2015/2016	WKIS	0.09	0.10	0.07	0.05	0.01	0.07	0.04	0.03	0.04	0.02	0.51
	BIS	0.11	0.08	0.06	0.09	0.01	0.07	0.06	0.04	0.05	0.02	0.60
	AIS	0.04	0.05	0.04	0.09	0.02	0.07	0.05	0.06	0.03	0.03	0.47
2016/2017	WKIS	0.08	0.11	0.08	0.05	0.01	0.07	0.05	0.03	0.04	0.02	0.54
	BIS	0.11	0.11	0.09	0.09	0.01	0.07	0.06	0.02	0.04	0.02	0.62

IS—Irrigation scheme; AIS—Ahero irrigation scheme; WKIS—WKano irrigation scheme; BIS—Bunyala irrigation scheme; RIS—relative irrigation supply; WSIA—irrigation supply per unit irrigated area; WSCA—irrigation supply per unit command area; WFC—water fee collection; ARIWS—annual revenue per unit irrigation supply; OIA—output per unit irrigated area; OCA—output per unit command area; OIS—output per unit irrigation supply; OCWD—output per unit water consumed; OWS—output per unit water supply.

Comparison of the trend in irrigation scheme performance of each scheme is presented in Figure 2.

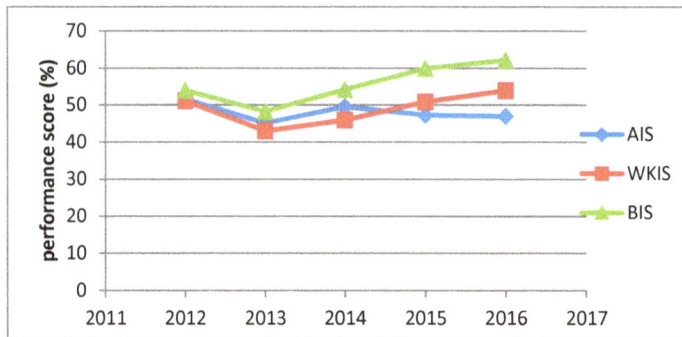

Figure 2. Comparison of performance score.

The overall performance score obtained was 45–52% in Ahero, 43–54% in West Kano and 48–62% in the Bunyala irrigation scheme. The average performance was 48%, 49% and 56% in the Ahero, West Kano and Bunyala irrigation schemes, respectively. The performance in all of the schemes was moderate. The performance in the West Kano and Bunyala irrigation schemes increased with time. The performance in the Ahero irrigation scheme was seen to be decreasing with time. The West Kano irrigation scheme experienced a fall in performance in 2014 due to the collapse of the Revolving Fund Committee which was mandated with the responsibility of production and marketing. The establishment of the production management structure, which created competition among the blocks in terms of production, increased performance from 2015. The sudden decline in performance in Bunyala in 2013 was due to hail stones that shattered mature rice crops in one of the phases (Muluwa phase 1). The reduction in the amount of water available due to drought led to a decrease in performance in the Ahero irrigation scheme in 2013. In 2013, there was strong sensitisation in the System of Rice Intensification (SRI) technology in Ahero. SRI involves changes in plants, water soil and nutrients management aimed at increasing productivity of rice under irrigation. Most farmers in the Ahero irrigation scheme adopted SRI, which led to an increase in performance from 45% in 2013 to 49% in 2014. A high performance of 83% [12] was obtained in the Samrat Ashok Sagar major irrigation project in India using a balanced score card method based on the Delphi technique. Agricultural productivity in India is highly enhanced by the government through artificial fixing of the minimum price of crops. The prices are therefore reasonably high, leading to the high economic value of crops. This is not the case in Kenya, where the price of rice produce is governed by market forces. In times of surplus, rice fetches low prices, reducing its economic value. This contributes greatly to low agricultural productivity performance, leading to low overall performance of irrigation schemes. Zema and Nicotra [42] used PCA to identify areas of weakness in seven Water Users' Association (WUA's) in Calabra, Southern Italy. The Ionio Catanzarese (ICZ) WUA was ranked as the best performing with a quality index of 4470, while the Basso Ionio Reggino (BIRC) was found to be the least performing with a quality index of −1410. BIRC was found to have a weakness in both system operation performance and financial management. Lowering water prices was found to be the solution to improving performance of BRIC WUA's in Calabra, Southern Italy [42].

4. Conclusions

The combination of benchmarking and Principal Component Analysis forms a powerful tool for evaluating the efficiency of irrigation schemes. The quantitative evaluation of performance of three rice irrigation schemes in western Kenya using a set of benchmarking indicators revealed the areas that needed improvement. Analysis of water supply indicators shows that, the water supplied by the irrigation schemes is sufficient to meet crop water demands. The irrigation schemes have low water use efficiency. In terms of financial performance, the irrigation schemes are not financially self-sufficient. The water fee charged was not sufficient to pay for the cost of irrigation. Land and water productivity in western Kenyan rice irrigation schemes was found to be generally good. Computation of a single performance score using performance indicators and principal component analysis enabled ranking of the irrigation schemes. The Bunyala irrigation scheme was found to be the best performing scheme, whereas the Ahero irrigation scheme was the least performing in the region. The overall performance of public rice irrigation schemes in western Kenya is average. Operation and management measures should be put in place to improve performance. The schemes need to adopt a systematic routine data collection and management to aid in the monitoring and evaluation of performance. Stakeholders and scheme managers can use this information to reformulate policies and strategies to enhance performance of public rice irrigation schemes in Kenya.

Author Contributions: F.M.M. designed the methodology, analysed data, interpreted the results and prepared the manuscript. P.G.H. conceived the research idea. J.M.R. assisted in obtaining the datasets. P.G.H. and J.M.R. guided in analysis and correction of the initial draft.

Funding: This research was funded by the African Union Commission (AUC) and Japan International Cooperation Agency (JICA).

Acknowledgments: The authors are grateful to National Irrigation Board (NIB) for granting access to western Kenyan irrigation schemes. We are also grateful to the Kenya Meteorological Department (KMD) and the management of the Ahero, Bunyala and West Kano irrigation schemes for providing historical data sets for this study.

Conflicts of Interest: The authors have no conflict of interest.

References

1. Government of Kenya. *National Water Master Plan 2030 in the Republic of Kenya Final Report Volume -v Sectoral Report (2/3)*; Government of Kenya: Nairobi, Kenya, 2013.
2. Ngenoh, E.; Kirui, L.K.; Mutai, B.K.; Maina, M.C.; Koech, W.; Victoria, L. Economic Determinants of the Performance of Public Irrigation Schemes in Kenya. *J. Dev. Agric. Econ.* **2015**, *7*, 344–352. [CrossRef]
3. Gitonga, K. *Grain and Feed Annual 2017 Kenya Corn, Wheat and Rice Report*; USDA Foreign Agricultural Service: Washington, DC, USA, 2017.
4. Ministry of Water and Irrigation. *Irrigation Sub-Sector MTP111 Zero Draft (2)*; Ministry of Water and Irrigation: Nairobi, Kenya, 2017.
5. Evans, A.A.; Florence, N.O.; Eucabeth, B.O.M. Production and Marketing of Rice in Kenya: Challenges and Opportunities. *J. Dev. Agric. Econ.* **2018**, *10*, 64–70. [CrossRef]
6. Kenya National Bureau of Statistics (KNBS). *Economic Survey of Kenya*; KNBS: Nairobi, Kenya, 2016.
7. Kenya National Bureau of Statistics (KNBS). *Kenya Population Census Report*; KNBS: Nairobi, Kenya, 2010.
8. Krhoda, G.O. *Kenya National Water Development Report*; UN-Water: Geneva, Switzerland, 2006; pp. 1–244.
9. Ngigi, S.N. Review of Irrigation Development in Kenya. In *The Changing Face of Irrigation in Kenya: Opportunities for Anticipating Change in Eastern and Southern Africa*; Blank, H.G., Mutero, C.M., Murray-Rust, H., Eds.; International Water Management Institute (IWMI): Colombo, Sri Lanka, 2002; Volume XIV, pp. 35–54.
10. Malano, H.; Burton, M.; Makin, I. *Guidelines for Benchmarking Performance in the Irrigation and Drainage Sector*, 1st ed.; International Programme for Technology and Research in Irrigation and Drainage (IPTRID) Knowledge Synthesis Report No. 5; FAO: Rome, Italy, 2001; Volume 53.
11. Malano, H.; Hungspreug, S.; Plantey, J.; Bos, M.G.; Vlotman, W.F.; Molden, D.; Burton, M. *Benchmarking of Irrigation and Drainage Projects*; International Commission on Irrigation and Drainage (ICID): New Delhi, India, 2004.
12. Phadnis, S.S.; Kulshrestha, M. Evaluation for Measuring Irrigation Service Performance Using a Scorecard Framework. *Irrig. Drain.* **2013**, *62*, 181–192. [CrossRef]
13. Renault, D.; Facon, T.; Wahaj, R. *Modernizing Irrigation Management: The MASSCOTE Approach-Mapping System and Services for Canal Operation Techniques*; Food and Agriculture Organization: Rome, Italy, 2007; Volume 63.

14. Burt, C.M.; Styles, S.W. Rapid Appraisal Process (RAP) and Benchmarking. *Bioresour. Agric. Eng.* **2001**, 32–80.

15. Molden, D.; Sakthivadivel, R.; Perry, C.J.; De Fraiture, C.; Kloezen, W.H. *Indicators for Comparing Performance of Irrigated Agricultural Systems*; IWMI: Colombo, Sri Lanka, 1998; ISBN 9290903562.

16. Rodríguez Díaz, J.A.; Camacho-Poyato, E.; López-Luque, R.; Perez-Urrestarazu, L. Benchmarking and Multivariate Data Analysis Techniques for Improving the Efficiency of Irrigation Districts: An Application in Spain. *Agric. Syst.* **2008**, *96*, 250–259. [CrossRef]

17. Córcoles, J.I.; de Juan, J.A.; Ortega, J.F.; Tarjuelo, J.M.; Moreno, M.A. Evaluation of Irrigation Systems by Using Benchmarking Techniques. *J. Irrig. Drain. Eng.* **2012**, *138*, 225–234. [CrossRef]

18. Borgia, C.; García-Bolaños, M.; Li, T.; Gómez-Macpherson, H.; Comas, J.; Connor, D.; Mateos, L. Benchmarking for Performance Assessment of Small and Large Irrigation Schemes along the Senegal Valley in Mauritania. *Agric. Water Manag.* **2013**, *121*, 19–26. [CrossRef]

19. Jamilah, M.; Zakaria, A.; Md. Shakaff, A.Y.; Idayu, N.; Hamid, H.; Subari, N.; Mohamad, J. Principal Component Analysis—A Realization of Classification Success in Multi Sensor Data Fusion. In *Principal Component Analysis—Engineering Applications*; InTech: Philadelphia, PA, USA, 2012; pp. 1–25. [CrossRef]

20. Paul, L.C.; Suman, A.A.; Sultan, N. Methodological Analysis of Principal Component Analysis (PCA) Method. *Int. J. Comput. Eng. Manag.* **2013**, *16*, 32–38.

21. OECD. *Handbook on Constructing Composite Indicators: Methodology and User Guide*; OECD Publishing: Paris, France, 2008.

22. Kipkorir, E.C. Rice productivity, Water and Sanitation Baseline Survey Report Western Kenya Rice Irrigation Schemes. Available online: https://www.researchgate.net/publication/322686337_Rice_productivity_Water_and_Sanitation_Baseline_Survey_Report_Western_Kenya_Rice_Irrigation_Schemes (accessed on 11 October 2018).

23. Balderama, O.F.; Bareng, J.L.R.; Alejo, L.A. Benchmarking for Performance Assessment of Irrigation Schemes: The Case of National Irrigation Systems and Small Water Impounding Projects in Cagayan River Basin. In Proceedings of the International Conference of Agricultural Engineering, Zurich, Switzerland, 6–10 July 2014.

24. Brouwer, C.; Heibloem, M. Irrigation Water Management: Irrigation Water Needs. *Train. Man.* **1986**, *3*, 225–240.

25. Bastiaanssen, W.; Perry, C. *Agricultural Water Use and Water Productivity in the Large Scale Irrigation (LSI) Schemes of the Nile Basin*; Nile Information System: Entebbe, Uganda, 2009.

26. Bastiaanssen, W.G.M.; Steduto, P. The Water Productivity Score (WPS) at Global and Regional Level: Methodology and First Results from Remote Sensing Measurements of Wheat, Rice and Maize. *Sci. Total Environ.* **2017**, *575*, 595–611. [CrossRef] [PubMed]

27. Omondi, S.O. *Economic Valuatio of Irrigation Water in Ahero Irrigation Scheme*; University of Nairobi: Nairobi, Kenya, 2014.

28. Bos, M.G.; Burton, M.A.; Molden, D.J. *Irrigation and Drainage Performance Assessment Practical Guidelines*; CABI Publishing: Wallingford, UK, 2005; ISBN 0851999670.

29. Kuscu, H.; Bölüktepe, F.E.; Demir, A.O. Evaluation Performance of Irrigation Water Management: A Case Study of Karacabey Irrigation Scheme in Turkey. *Int. J. Agric. Sci.* **2015**, *5*, 824–831.

30. Kuşçu, H. Benchmarking Performance Assessment of Irrigation Water Management in a River Basin: Case Study of the Susurluk River Basin, Turkey. *Afr. J. Bus. Manag.* **2012**, *6*, 2848–2859. [CrossRef]

31. Government of Maharashtra. *Report on Benchmarking of Irrigation Projects in Maharashtra*; Government of Maharashtra: Mumbai, India, 2005.

32. Brouwer, C.; Prins, K.; Heibloem, M. Irrigation Water Management: Irrigation Scheduling. *Train. Man.* **1989**, *4*, 66.

33. Zema, D.A.; Nicotra, A.; Zimbone, S.M. Diagnosis and Improvement of the Collective Irrigation and Drainage Services in Water Users' Associations of Calabria (Southern Italy). *Irrig. Drain.* **2018**. [CrossRef]

34. Brouwer, C.; Prins, K.; Kay, M.; Heibloem, M. Irrigation Water Management: Irrigation Methods. *Train. Man.* **1998**, *5*, 140.

35. Cin, S.; Çakmak, B. Assessment of Irrigation Performance in Başören Irrigation Cooperative Area of Beypazarı, Ankara. *J. Agric. Fac. Gaziosmanpaşa Univ.* **2017**, *34*, 10–19. [CrossRef]

36. De Alwis, S.M.D.L.K.; Wijesekara, N.T.S. Comparison of Performance Assessment Indicators for Evaluation of Irrigation Scheme Performances in Sri Lanka. *Engineer* **2011**, *44*, 39–50. [CrossRef]

37. Ghazalli, M.A. Benchmarking of Irrigation Projects in Malaysia: Initial Implementation Stages and Preliminary Results. *Irrig. Drain.* **2004**, *53*, 195–212. [CrossRef]

38. Shenkut, A. Performance Assessment Irrigation Schemes According to Comparative Indicators: A Case Study of Shina-Hamusit and Selamko, Ethiopia. *Int. J. Sci. Res. Publ.* **2015**, *5*, 451–460.

39. Mchele, A.R. Summary for Policymakers. In *Climate Change 2013—The Physical Science Basis*; Intergovernmental Panel on Climate Change, Ed.; Cambridge University Press: Cambridge, UK, 2011; Volume 53, pp. 1–30.

40. Bumbudsanpharoke, W.; Prajamwong, S. Performance Assessment for Irrigation Water Management: Case Study of the Great Chao Phraya Irrigation Scheme. *Irrig. Drain.* **2015**, *64*, 205–214. [CrossRef]

41. Kalantari, K.H. Processing and Analysis of Economic Data in Social Research. *Publ. Consult. Eng. Landsc. Des. Tehran* **2008**, *3*, 110–122.

42. Zema, D.A.; Nicotra, A.; Tamburino, V. Performance Assessment of Collective Irrigation in Water Users' Associations of Calabria (Southern Italy). *Irrig. Drain.* **2015**, *64*, 314–325. [CrossRef]

![agriculture logo] *agriculture*

MDPI

Article

Assessing Heat Management Practices in High Tunnels to Improve the Production of Romaine Lettuce

Muzi Zheng [1,*], Brian Leib [1], David Butler [2], Wesley Wright [1], Paul Ayers [1], Douglas Hayes [1] and Amir Haghverdi [3]

[1] Department of Biosystems Engineering and Soil Science, University of Tennessee, 2506 E.J. Chapman Drive, Knoxville, TN 37996-4531, USA; bleib@utk.edu (B.L.); wright1@utk.edu (W.W. & P.A.); dhayes1@utk.edu (D.H.)

[2] Department of Plant Sciences, University of Tennessee, 2431 Joe Johnson Dr., Knoxville, TN 37996-4531, USA; dbutler@utk.edu

[3] Department of Environmental Sciences, University of California, Riverside, 900 University Avenue, Riverside, CA 92521, USA; amirh@ucr.edu

* Correspondence: mzheng3@vols.utk.edu

Received: 2 August 2019; Accepted: 9 September 2019; Published: 14 September 2019

Abstract: A three-year experiment evaluated the beneficial effects of independent and combined practices on thermal conditions inside high tunnels (HTs), and further investigated the temperature impacts on lettuce production. Specific practices included mulching (polyethylene and biodegradable plastic films, and vegetative), row covers, cover crops, and irrigation with collected rainwater or city water. The study conducted in eastern Tennessee was a randomized complete block split-split plot design (RCBD) with three HTs used as replicates to determine fall lettuce weight (g/plant) and lettuce survival (#/plot), and the changes in soil and air temperature. The black and clear plastic mulches worked best for increasing plant weight, but when compared to the bare ground, the higher soil temperature from the plastics may have caused a significant reduction in lettuce plants per plot. Moreover, the biodegradable mulch did not generate as much soil warming as black polyethylene, yet total lettuce marketable yield was statistically similar to that for the latter mulch treatment; while the white spunbond reduced plant weight when compared with black plastic. Also, row covers provided an increased nighttime air temperature that increased soil temperature, hence significantly increased lettuce production. Cover crops reduced lettuce yield, but increased soil temperatures. Additionally, irrigation using city water warmed the soil and provided more nutrients for increased lettuce production over rainwater irrigation.

Keywords: cover crop; lettuce production; irrigation; mulch; row cover; temperature variations

---

1. Introduction

High tunnels (HTs) are simple, plastic covered, greenhouse-like structures that do not utilize heaters or ventilation fans. Even without heaters or fans, HTs allow producers to lengthen the growing season and protect plants from extreme weather conditions (e.g., hail, frost, or strong wind), hence increasing the profitability and sustainability of organic farms [1]. Compared to crop production in open fields, other advantages of HTs highlighted by Lamont et al. [2] also include: (1) preventing rainfall from wetting and splashing soil onto fruit and foliage, thus creating cleaner products with less disease; (2) increasing water-use efficiency; (3) improving crop environmental conditions; and (4) reducing soil compaction and insect invasion. Additionally, the benefits of less electricity consumption, lower startup costs, and fewer maintenance efforts make HT systems more competitive than standard greenhouses [3].

High-valued crops, such as lettuce, tomatoes, and other leafy greens, produced inside HT systems can also compensate for the increased cost of HT systems when compared to open field production. Studies also show that the microclimates inside HTs can be modified so that the growing season can be lengthen from 1 to 4 weeks in spring and 2 to 8 weeks in fall [4]. Therefore, HTs are currently being adopted by small- and mid-scale producers in order to take advantages of market seasonality with higher profits [5,6].

However, there are several limitations for the sustainability of HT systems. First, since HTs block rainfall, ground/well or treated water needs to be used for crop irrigation. As well, lettuce is considered moderately sensitive to irrigation salinity, since the salts in soil and water might cause the retardation of crop growth, reduction of lettuce head formation, or even the necrotic lesions on the leaf margins [7]. To combat this sensitivity, rainwater can be collected and reserved in black polyethylene tanks, and then used inside HTs through drip irrigation. A rain harvesting system (RHS) is considered beneficial in regions where treated water is inadequate and costly, and helps in reducing local flooding in some low-lying areas. This study investigates additional benefits of RHS when black polyethylene tanks are used; black tanks can capture solar energy and warm up enclosed rainwater, and this warmer water can warm the soil during drip irrigation.

Secondly, the poor insulation properties of the plastic-covered HT structure do not provide significant heat retention at night under cold climates. Without a favorable growth microclimate, crop performances inside HTs might be inhibited or even terminated. Accordingly, surface mulch and row covers are practices that could help retain heat while providing other advantages, such as reducing weed competition, alleviating soil crusting, modifying the radiation budget of the soil surface around the plants, and conserving water by inhibiting evaporation from the soil surface [8]. A two-year lettuce experiment in the open field found that colored polyethylene mulches (clear, white, and black) significantly increased the overall lettuce production by 33%, 40%, and 39%, respectively, compared to bare ground [9]. Additionally, for other high-value crops, a study showed that black polyethylene mulch combined with an HT system significantly increased overall pepper production in regard to crop height and leaf area, while in the same system, clear plastic mulch significantly raised ground temperature, thus decreasing the number of days to the first flower when compared with black polyethylene mulch and bare ground [10]. As well, biodegradable mulch can eliminate plastic disposal issues, including the labor cost of disposal [11,12]. There is limited research on how biodegradable mulches within HT systems affect soil temperature and crop productivity. But in open field tests, outdoor use of biodegradable mulches had no significant effect on overall pepper yield and quality in Australia; although, it raised soil temperature slightly compared with paper mulch [13]. A study from Moreno [14] also indicated that biodegradable mulches did not significantly increase tomato production in Central Spain, but Miles et al. [15] observed improved tomato production using biodegradable mulches when compared with bare ground in northwestern Washington, USA. Since lettuce production could be limited from weed competition and seedling establishment problems, appropriate polyethylene and biodegradable mulches might reduce the costs of weed management, while minimizing root damage, also retaining favorable soil temperatures for lettuce growth and development [16]. Similar to surface mulches, floating row covers are considered as additional thermal protection for crops from wind and frost inside a HT. There have been many studies showing that the combination of mulching and row covers under field conditions significantly increased soil and air temperatures for favorable crop growth conditions [17]. However, some studies indicated that although HT air temperature was slightly higher than outside, lettuce development was still slow, since the air temperature around the plant's full height was not optimal [18]. Therefore, more investigation on the impacts of row cover and colored plastic/biodegradable mulches under HTs are needed to maximize heat retention.

The last limitation is related to the decomposition of soil organic matter due to continuous cropping in HTs, which may negatively influence nutrient/water holding capacity in soil. The common solution of adding organic supplements to the soil requires energy at each step as long as the supplements need to be collected and delivered to HTs. Accordingly, an effective practice is to pre-plant a legume cover crop which has the ability to fix nitrogen, supply crop nutrients and improve soil structure in HT systems. This study investigated incorporating cover crop residues into the ground and leaving the residues on the soil surface (vegetative mulch). Teasdale and Daughtry [19] found that cover-crops have positive effects on weed control and Liebman and Davis [20] observed reduced maximum-soil-temperature under vegetative mulches in the daytime, along with the benefit of increased summer production. Even though cover crops provide many benefits, there is limited research on the effects of cover crop residues incorporated into the ground in combination with polyethylene mulches covering the soil surface, and their effects on thermal protection in HT systems.

Overall, this study aims to provide recommendations to small- and mid-scale producers on how to improve thermal protection with low input sustainable practices to improve lettuce production in HTs. Rain water, surface mulch, row covers, and cover crops have many benefits, but this study will focus on the benefits to heat management. Zheng et al. [21]'s study showed how thermal energy conservation benefited spring pepper production, while this study aims to assess the yield performance of fall Romaine lettuce using the same thermal energy conservation practices.

2. Material and Methods

2.1. The Gothic Type High Tunnel

Three experimental high tunnels were used in the experiment, that were orientated N–S, and located at the University of Tennessee's Organic Crop Unit in Knoxville, TN in the United States (latitude 35.88° N, longitude 83.93° W, and elevation 252 m). These HTs have peaked roofs, vertical side walls and the dimension of the structures are 9 m in width and 15 m in length (Figure 1). Each end door covers an opening which is 2.45 m tall and 3.35 m wide, and side curtains run the entire length of the HT. The end doors were opened almost every day to provide natural ventilation while the side-curtains were only opened during warm weather when more air was need for proper ventilation. In cold periods, doors and side curtains were closed at night to preserve thermal energy, and only the end doors were opened in the daytime to reduce accumulated heat and humidity inside HTs. Rainwater was collected and stored in black polyethylene tanks and was delivered using gravity pressure and solar power. Rainwater storage tanks in the first HT were elevated with cinder blocks to provide gravity pressure for irrigation with drip tubing, while the other two HTs utilized a solar powered pump to deliver rainwater via drip tapes with 10 psi of pressure [21].

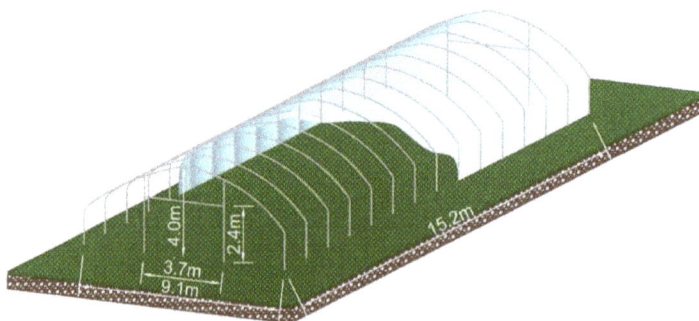

Figure 1. Isometric view of a high tunnel with the covering cut away for clarity) [21].

Romaine Lettuce ('coastal star') was grown during the fall seasons of 2011, 2013, and 2014. Specifically, five different surface mulches with a bare soil control, and two sources of irrigation water were applied in fall 2011. Thus, there were six beds in each HT, and these six different treatments were randomly arranged in different beds. Figure 2a shows that mulch treatment was laid out in six rows, including: (1) white biodegradable spunbond (spun), (2) vegetative mulch from a leguminous cover crop (veg), (3) bare ground (bare), (4) black polyethylene film (black), (5) biodegradable brown paper (paper), and (6) black biodegradable biobag (biobag). The dimension of each bed was 12.19 m in length and 0.91 m in width with double rows of lettuce planted in each bed. Double planting rows were placed 0.3 m apart and 0.46 m apart between lettuce transplants. There were 48 plants per row and rows were divided into four plots. Lettuce was transplanted on 30 September 2011 in half of each house (snapdragon flowers were added to the other half of the HT for a different research project), and then harvested on 7 December 2011. Next, in the fall of 2013, there were five surface mulches and a bare control, with or without a row cover. Figure 2b shows that the mulch distribution was: (1) black polyethylene film (black), (2) vegetative mulch from a cover crop (veg), (3) bare ground (bare), (4) black polyethylene film with a cover crop (blackCC), (5) clear polyethylene film with a cover crop (clearCC), and (6) clear polyethylene film (clear). Once inside temperature fell below approximately 7 °C, row covers were applied to half of the crop during the night. Lettuce was transplanted on 2 October 2013 and harvested on 4 December 2013. Finally, in the fall of 2014, a total of 12 treatments were used, including two water resources (rainwater and city water), two surface mulches (black and blackCC) and a bare ground control, with or without a row cover (Figure 2c). Lettuce was transplanted in each half of the HT on 29 September 2014 and harvested on 5 December 2014.

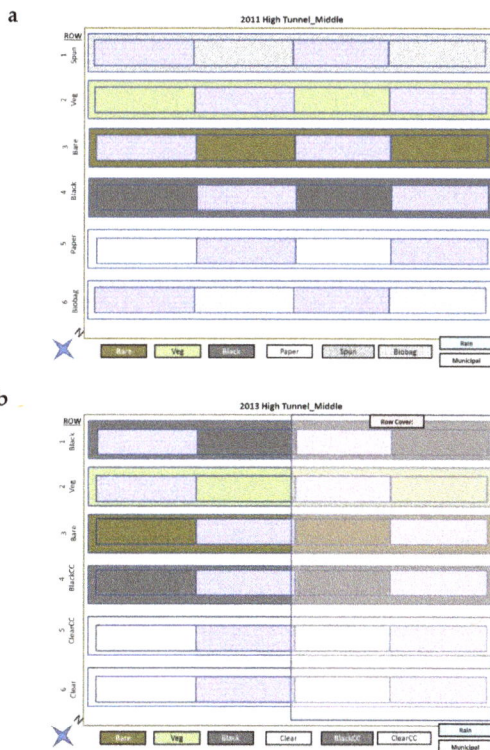

Figure 2. *Cont.*

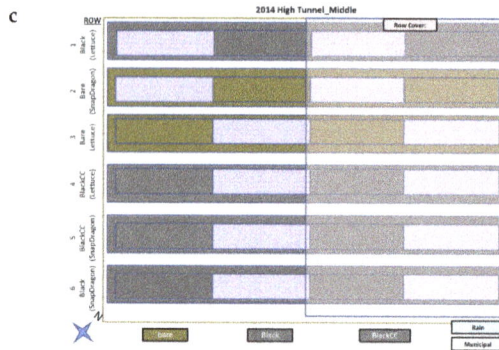

Figure 2. The lettuce production of high tunnels (HTs) in 2011, 2013, and 2014 followed a split-split-plot design.

To determine statistical differences, the experiment was a randomized complete block design (RCBD), based on a split-split-plot sub-design for lettuce production and temperature variations. In addition, soil at the experimental site was a Dewey silt loam. Lettuce yield was determined by plant weight (g/plant) and plants per plot (#/plot). Climatic analysis were divided into two time periods, including early fall (EF) from planting through October and late fall (LF) from November to early December, and statistical analyses of lettuce yield values and temperatures variations were performed by SAS statistical software (SAS Institute Inc., Cary, NC, USA).

2.2. Climatic Monitoring and Instrumentation

Two meteorological stations were used to monitor conditions with one placed outside and another placed inside the high tunnels [21]. The outside station was 6.0 m away from the middle HT and set to 4.0 m above the ground. The inside station was located at 1.5 m above the ground at the center of the middle experimental HT. The sensors used for each station included a pyranometer for solar radiation, a cup anemometer and wind vane for wind speed and direction, along with a capacitive chip and platinum resistance thermometer, for relative humidity and air temperature, respectively. Additionally, there was a total of 15 thermistors applied to measure inside air temperature at 1.2 m above the ground surface. An array of 12 thermistors to monitor air temperatures at the canopy level were located 30 cm above the ground surface. Half of these thermistors were under the row cover and half were outside the row cover. Moreover, soil temperature was measured in each combination for the mulch, row cover, and water using 24 thermistors placed 10 cm into the ground. Additionally, the water temperatures were measured at the bottom of polyethylene storage tanks and in the supply line of city water. All the thermistors described above were potted in solar reflecting white epoxy and all the sensors were measured every 15 s and recorded every hour by a CR1000 datalogger (Campbell Scientific, Inc.: Logan, Utah, USA). The specifications for the sensors are listed in Table 1.

Table 1. Environmental parameters measured during the experiments [21].

Measured Parameters	Sensors	Range and Accuracy	Location
Solar radiation Q_{solar} (Wm^{-2})	LI200X Pyranometer (LI-COR.INC)	0 to 3000 Wm^{-2} with ±5%	Inside and outside weather station
Air temperature T_i (K) Relative humidity H_R (%)	HMP60 Probe (Vaisala.INC)	−40 to 60 °C with ±0.6% ±3% RH over 0 to 90%, ±5% RH over 90 to 100%	Inside and outside weather station
Wind speed U (ms^{-1}) Wind direction (°)	R.M.Young Wind Sentry (Campbell Sci. Inc)	0 to 50ms^{-1} with ±1% 0 to 360° with ±5°	Outside weather station
Air temperature (°C)	Thermistor (PS104J2) [†]	±0.1 °C with 0.026 °C	An array of 15 thermistors all over the inside of middle HT located 1.5 m above the ground
Canopy temperature (°C)	Thermistor (PS104J2) [†]	±0.1 °C with 0.026 °C	12 thermistors located 30 cm above the ground between crop rows
Soil temperature (°C)	Thermistor (PS104J2) [†]	±0.1 °C with 0.026 °C	24 thermistors located 10.16 cm below the ground
Water temperature (°C)	Thermistor (PS104J2) [†]	±0.1 °C with 0.026 °C	Polyethylene storage tank bottom for rainwater and supply line of city water

[†] means the thermistors are made in our lab, which are non-linear sensors following a polynomial response to temperature that is defined by the Steinhart–Hart equation.

3. Results and Discussion

The HT affected inside microclimate in terms of solar radiation, air and soil temperature, and relative humidity and wind speed. There were 1380 observations of weather data measured over three falls of 2011, 2013, and 2014. Generally, 9% of total solar radiation was reflected by the HT's polyethylene covering (dataset in 2011 was disregarded due to sensor failure). During the EF (early fall) with the end-doors and side-curtains fully opened, air temperatures inside the HT were consistently higher by 0.5–6.0 °C compared to outside; while in LF (late fall) with the side curtains rolled down, the inside maximum air temperature during the day was from 1 to 11 °C higher when compared to the outdoor conditions. Moreover, outside and inside thermistor readings indicated that night time air temperatures inside the HT were 0.6–2.0 °C higher than outdoor temperatures, and the average nighttime soil temperature inside the HT was 1–3 °C higher than outside over the span of the three falls [21].

Statistically, mulch treatments were considered the main contributing factor to the lettuce productivity in the falls of 2011, 2013, and 2014 ($p < 0.001$) (Figure 3). In 2013 and 2014, black consistently produced a higher yield than bare, with an increase of 71% (2013) and 37% (2014) in g/plant. However, no significant difference was found in lettuce yield between using black and bare plots in 2011. Clear mulch provided the maximum production in g/plant in 2013 but with no significant difference compared to the black mulch treatment. In addition, Table 2 shows that the increased crop yield using black was related to its soil-warming ability when compared with bare. Although some temperature values were missing due to sensor failures in LF 2011, black had a positive impact on soil temperature, which was constantly higher than the bare treatment by 0.8–1.3 °C (2011), 0.6–0.9 °C (2013), and 0.9–1.3 °C (2014). The soil temperature rise under black mulch can promote faster lettuce growth and development, although the lettuce survival (#/plot) with black was reduced approximately 20% when compared to bare in 2013. The soil-warming ability underneath clear was even greater than that under black, by as much as 1.0 °C, but the overall lettuce yield did not improve significantly in spring 2013. After transplanting in EF, soil temperatures were warmer by 1.4 °C (day) and 2.1 °C (night) for clear, and 0.6 °C and 0.9 °C for black when compared to the bare. The higher measured soil temperatures under the mulches do not adequately reflect, the extremely hot air emitted from the transplant holes or the high temperature of the mulch itself before the crop canopy can shade the plastic mulches. These higher temperatures seem beneficial to lettuce growth (lettuce weights per plant) but they may not be beneficial to lettuce survival at transplant, since the plants per plot were significantly reduced in clear (10%) and black (20%) plots compared with the bare.

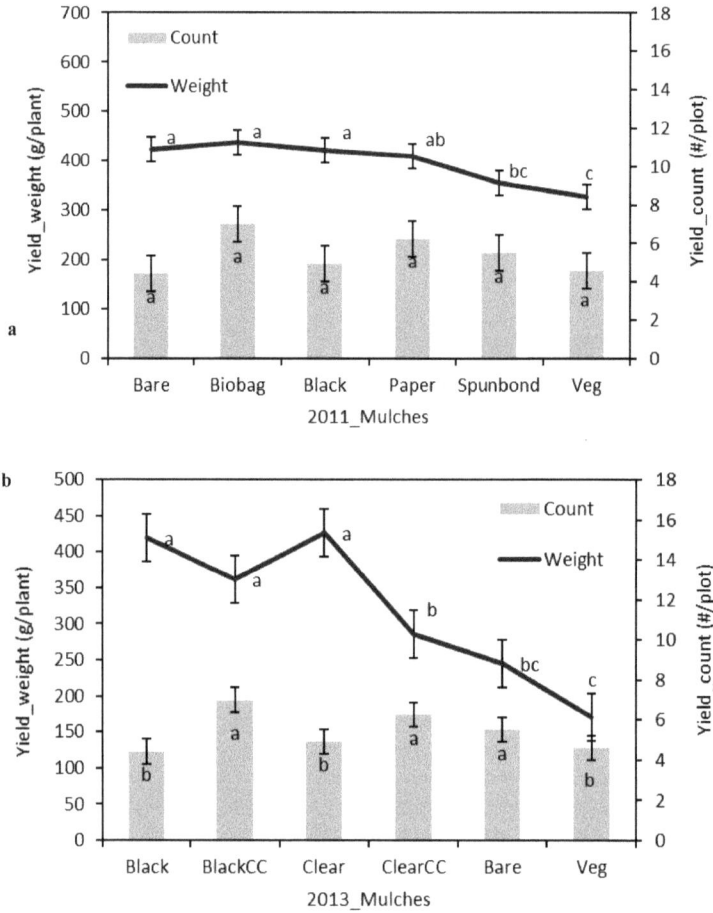

Figure 3. *Cont.*

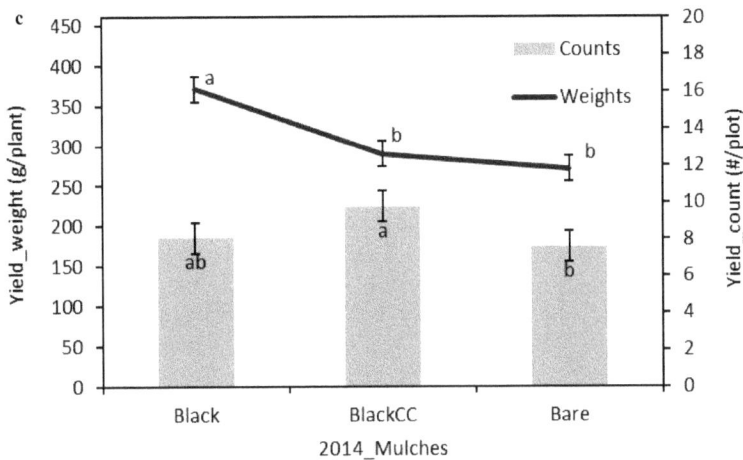

Figure 3. Marketable lettuce weight per plant and lettuce numbers per plot under the effects of different mulch over the falls of the years 2011 (**a**), 2013 (**b**), and 2014 (**c**) in HTs. Values followed by the same letter are not significant different according to Fisher's LSD (*p* = 0.1).

Table 2. Marketable lettuce weight per plant and lettuce numbers per plot, and average soil temperatures at 10 cm depth under different mulch treatments during the day and night in the early and late falls of 2011, 2013, and 2014.

Year	Mulch	Lettuce Yield		Soil Temp_Day		Soil Temp_Night	
		Weight	Count	EF[†]	LF[†]	EF[†]	LF[†]
		(g/plant)	(#/plot)	(°C)		(°C)	
	Black	421.6 a	7.7 a	18.22 a	-	14.42 a	15.39 a
	Biobag	436.9 a	8.3 a	17.12 b	-	13.26 c	14.17 c
2011	paper	409.1 ab	8.7 a	16.68 bc	-	-	-
	Spunbond	355.7 bc	9.3 a	16.21 c	-	13.01 d	13.99 d
	Bare	424.2 a	9.2 a	17.06 b	-	13.59 b	14.13 cd
	Veg	327.5 c	9.2 a	17.06 b	-	12.96 d	13.52 d
	Sig.level	*0.0311*	*0.73*	*0.0115*	-	*<0.0001*	*<0.0001*
	Black	419.4 a	4.4 b	19.97 c	14.12 b	19.74 c	14.31 e
	BlackCC	361.4 a	7.0 a	20.08 c	14.18 b	20.52 b	14.66 b
	Clear	425.8 a	4.9 b	20.80 b	14.20 b	20.86 b	14.66 b
2013	ClearCC	285.6 b	6.3 a	21.47 a	14.69 a	21.81 a	15.02 a
	Bare	244.9 bc	5.5 a	19.41 d	13.26 d	18.82 d	13.43 e
	Veg	170.4 c	4.6 b	19.22 e	13.01 d	18.36 d	13.28 e
	Sig.level	*<0.0001*	*<0.0001*	*<0.0001*	*<0.0001*	*<0.0001*	*<0.0001*
	Black	371.1 a	8.1 ab	18.38 a	13.07 a	19.48 a	13.86 a
2014	BlackCC	290.0 b	9.8 a	18.24 b	12.97 a	19.32 a	13.76 a
	Bare	271.2 b	7.6 b	17.30 c	12.12 b	18.16 b	12.86 b
	Sig.level	*<0.0001*	*0.1765*	*<0.0001*	*<0.0001*	*<0.0001*	*<0.0001*

* EF and LF were characterized by diurnal soil temperature in specific early and late spring. Temperature values are daily means from 10:00 a.m. to 16:00 p.m. and nightly means from 21:00 p.m. to 8:00 a.m. Values followed by the same letter are not significantly different according to Fisher's LSD (*p* = 0.1).

Moreover, the application of the dark colored biobag and paper were not significantly different when compared to black for lettuce yield in 2011. However, white spunbond produced less marketable lettuce weight per plant by 13% than paper, but the difference in #/plot was not at a significant level. All of the degradable mulch plots provided significantly lower soil temperatures than black by 1.0–1.2 °C (biobag), 1.4–2.0 °C (spunbond), and approximately 1.6 °C (paper). The greater reduction of soil temperatures produced by spunbond (1.4–2.0 °C) compared to those of biobag and paper mulch may be the reason for significantly decreased lettuce production under spunbond when compared to black.

In additional, polyethylene mulch with cover crops (blackCC and clearCC) had the potential to save more lettuce heads in each plot by 21%–60% (#/plot). This may be because more air was able to penetrate through the cover crop residues, thus the thermal stress from the plastic mulch surface was able to be mitigated when the film edges contacted the lettuce transplants in the warm EF period. But the plastic mulch with cover crops significantly reduced marketable lettuce weight per plant (18% in blackCC and 33% in clearCC), when compared to non-cover crop treatments as an average for 2013 and 2014. Specifically, the average soil temperatures under blackCC in EF and LF of 2013 were around 0.3–0.8 °C warmer compared with black at night, whereas the temperature differences between clearCC and clear were 0.5–0.7 °C and 0.3–0.9 °C warmer during the day and night, respectively. Thus, although the cover crops had a higher soil-warming ability and saved more lettuce in each plot, limited nutrient availability caused by immobilization of nitrogen during the decomposition of cover crop residues may have inhibited the growth of each lettuce head, as the incorporated cover crop and residual soil nutrients served as the only fertilizer source. In comparison, veg always significantly provided the lowest lettuce production. It appears that the vegetative mulch shielded the transfer of solar energy to the soil and cooled the soil to a point where lettuce development was detrimentally delayed. Results confirmed that veg had a significant decrease of soil temperature (0.6 °C in 2011 and 0.3 °C in 2013) than that in bare. Therefore, veg treatment with the coolest soil temperature was working against a goal of making better use of solar energy to extend the growing season.

For the LF, additional thermal protection was provided by row covers for lettuce growth (Figure 4). Row cover played a significant role in increasing g/plant for 2014 ($p < 0.0001$). The lettuce weight per plant utilizing a row cover was improved approximately 36% as an average of all treatments (black, blackCC, and bare) when compared to all treatments without row covers. Moreover, positive interactions between the applications of mulches and row covers were found on yield ($p < 0.1$) in fall 2013. The black mulch with row cover provided the largest lettuce production and the row covers increased lettuce weight by 28% over black mulch alone. While some of the lower yielding plots had greater lettuce weight when a row cover was applied, bare and veg yielded significantly more lettuce weight, approximately 45% and 73% more lettuce weight per plant respectively, under row cover. Moreover, blackCC, clear, and clearCC with a row cover had the potential to improve the total marketable production (g/plant) compared with those without using row covers in fall 2013, but the improvement was not to a significant level. In addition, the row cover can significantly increase canopy and soil temperature by 0.8–2.3 °C and 0.1–0.9 °C at night, respectively. Although soil temperature increases from row covers interacted differently with the mulch treatments—0.4 °C in black, 0.6 °C in blackCC, 0.1 °C in clear, 0.8 °C in clearCC, 0.7 °C in bare, and 0.2 °C in veg—the row covers had a positive effect on lettuce production in all plots.

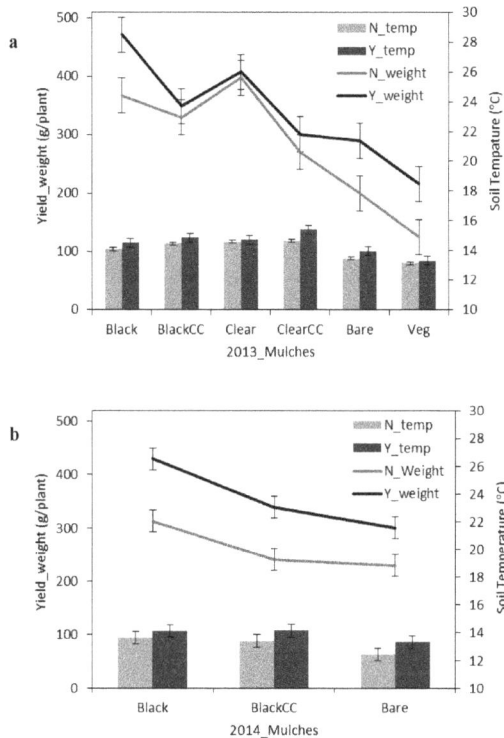

Figure 4. Interactions of mulch and row cover on lettuce yield (weight/plant), and soil temperature (°C) inside high tunnels during the late fall (LF) of 2013 and 2014. 'Y' represents the mulch with a row cover and 'N' means the mulch without a row cover.

Finally, lettuce yield was improved by using city water compared to rainwater applied with drip irrigation (Figure 5). As an average of all mulches, city water potentially increased the total lettuce weight per plant by 16% in 2011, 12% in 2013, and 8% in 2014, and also increased total lettuce numbers per plot by 13% in 2013 (Table 3). Although the numeric values of lettuce number per plot were potentially higher by using city water than rainwater, the differences were not at a significant level. In 2011, the biobag mulch and bare irrigated using city water significantly improved lettuce weights by 23% and 36%, respectively, compared to rainwater ($p < 0.0001$), although in subsequent years, city water did not significantly increase yield in the bare soil treatment. In 2013, black and veg mulch yield was significantly increased when city water was used for irrigation by 36% and 20% in lettuce weight per plant, respectively; and in 2014, black mulch with city water also significantly improved the lettuce production approximately by 17% (g/plant) over rainwater. The differences of lettuce yield using two water types were assumed to be explained by the impacts of water temperature on soil temperature. Figure 6 shows that the rainwater temperature from the solar tank was consistently cooler than that in gravity tanks. Additionally, in the EF of 2011, 2013, and 2014, the average temperature of rainwater stored in the solar tank was significantly higher than the city water in all years by 2.4 °C; but during the LF periods, the rainwater from solar tanks was significantly cooler than the city water by 2.2 °C in 2011, 1.3 °C in 2013 and 0.8 °C in 2014. Although water temperatures changed inconsistently between the EF and LF, soil temperatures responded in a similar manner over different mulches. Table 3 shows that city water generally produced warmer soil temperatures than rainwater during the entire fall seasons, with several exceptions in 2011 (biobag, spunbond, and bare), but all these variations cannot change the fact of greater lettuce production in city water plots (23% for biobag, 13% for spunbond, and 35% for bare). In summary, the temperature of city water was cooler than rainwater in EF and warmer during LF. However, the overall effect was for city water to warm the soil on average during each time period. As early fall progresses into late fall, rain water would be expected to cool in relation to city water, because the above ground tanks are exposed to less daily solar radiation and longer cooler nights, while soil temperature around the city water pipes lags behind the above ground heat loss of the rain water tank; i.e., soil temperature decline lags air temperature decline in the fall. The comparison of water temperature data does follow this expected pattern. However, the contradiction between EF water temperature and soil temperature differences is less easily explained. Perhaps the rainwater temperature was not representative of the water delivered for irrigation due to the sensor's location in the tank and temperature gradients within the tank caused by tank surface heating and cooling. In addition, the temperature transition from warmer rainwater to warmer city water would not occur precisely at the interface of EF to LF. There was most likely an intermixing of more and less heat transfer from the irrigation sources to the soil, in combination with air temperature's effects on soil temperature. Overall. the response of different mulches to irrigation water type shows that lettuce yield was improved by applying city water when compared to rainwater, and warmer soil temperatures appear to be an important factor where city water was applied. Another reason for increased lettuce production using city water may be related to water quality. Periodic water samples were analyzed and revealed the pH of city water was consistently around 6.4, while the rain water changed over time, but mainly was around at 6.7 to 7. Additionally, city water had more nutrients than rainwater: around 11 to 16 times higher levels of K, Mg, and Ca which may help promote lettuce growth and development.

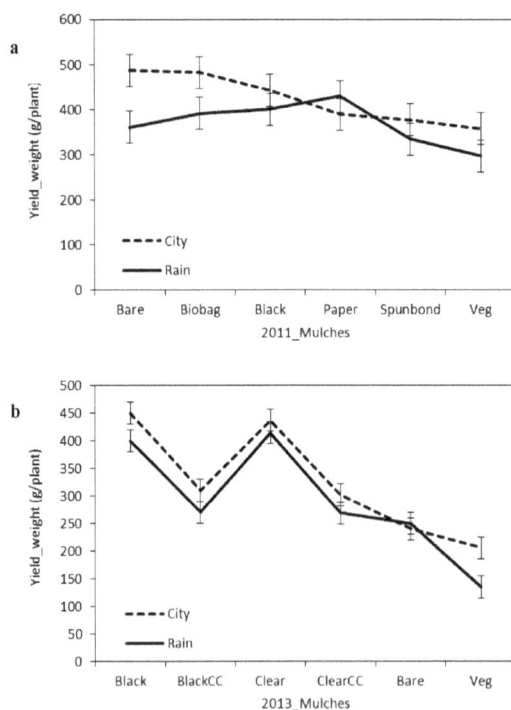

Figure 5. *Cont.*

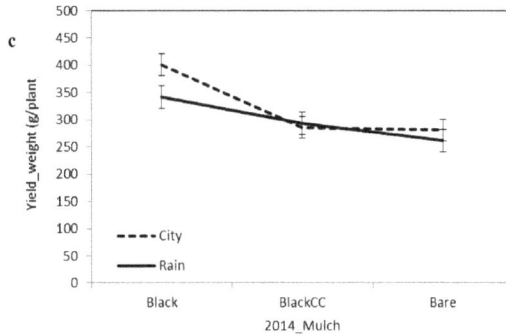

Figure 5. Marketable weight of lettuce (mean ± SE) under the effects of different irrigation water (rain and city) over three fall seasons of the years 2011 (**a**), 2013 (**b**), and 2014 (**c**) in high tunnels.

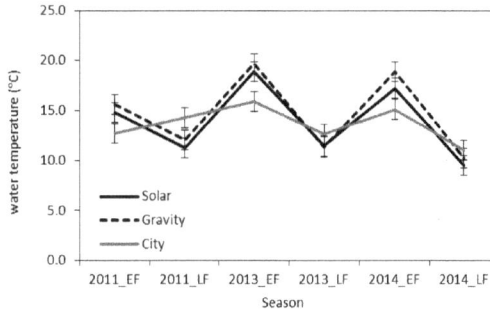

Figure 6. Fluctuations in average water temperature (mean ± SE) for three fall seasons of years 2011, 2013, and 2014. Gravity and Solar represent the rainwater inside the gravity and solar tanks, respectively, and City represents the municipal water from the underground pipes.

Table 3. Interactions of mulch and water on the lettuce yield and soil temperature inside high tunnels during all three falls of years 2011, 2013, and 2014.

Year	Mulch	Water	Lettuce Yield		Soil Temp_Day		Soil Temp_Night	
			Weight	Count	EF[†]	LF[†]	EF[†]	LF[†]
			(g/plant)	(#/plot)	(°C)		(°C)	
2011	Black	City	442.67 abc	10.00 a	18.37 a	-	14.44 a	15.45 a
		Rain	400.58 bcd	7.33 ab	18.07 ab	-	14.40 a	15.34 a
	Biobag	City	482.52 ab	8.67 ab	16.94 bcd	-	13.10 fg	13.83 e
		Rain	391.33 cd	8.00 ab	17.30 abc	-	13.42 def	14.52 bc
	Paper	City	388.67 cd	9.67 ab	16.80 bcd	-	13.48 cde	14.53 bc
		Rain	429.50 abc	9.00 ab	16.56 c	-	-	-
	Spunbond	City	377.33 cde	8.33 ab	15.90 d	-	12.87 g	13.77 e
		Rain	334.00 de	10.00 a	16.53 c	-	13.15 efg	14.21 d
	Bare	City	487.33 a	8.33 ab	17.11 bc	-	13.19 efg	13.79 e
		Rain	361.04 cde	7.0 b	17.01 bc	-	13.99 b	14.48 c
	Veg	City	357.69 cde	9.00 ab	17.34 abc	-	13.61 cd	14.72 b
		Rain	297.33 e	9.00 ab	16.78 bcd	-	13.78 bc	14.55 bc
	Mulch		*0.0311*	*0.73*	*0.0115*	-	*<0.0001*	*<0.0001*
	Water		*0.0158*	*0.38*	*0.8917*	-	*0.0018*	*<0.0001*
	Mulch × Water		*0.3177*	*0.59*	*0.5621*	-	*0.0678*	*<0.0001*

Table 3. *Cont.*

Year	Mulch	Water	Lettuce Yield		Soil Temp_Day		Soil Temp_Night	
			Weight	Count	EF[†]	LF[†]	EF[†]	LF[†]
			(g/plant)	(#/plot)	(°C)		(°C)	
2013	Black	City	450.0 a	5.17 bcd	20.15 c	14.28 c	20.04 cd	14.57 d
		Rain	400.0 a	3.67 d	19.79 d	13.97 d	19.43 ef	14.04 e
	BlackCC	City	310.3 abc	8.00 a	20.09 c	14.16 c	20.48 c	14.64 cd
		Rain	270.0 ab	6.00 cd	20.07 c	14.20 c	20.57 c	14.68 cd
	Clear	City	437.0 a	5.00 bcd	20.83 b	14.24 c	21.23 b	14.74 c
		Rain	414.7 a	4.8 bc	20.77 b	14.16 c	20.49 c	14.57 d
	ClearCC	City	302.2 bcd	6.17 bc	21.54 a	14.80 a	21.78 ab	15.10 a
		Rain	269.0 cd	6.33 bc	21.40 a	14.59 b	21.83 a	14.93 b
	Bare	City	239.7 de	6.50 ab	19.46 ef	13.48 e	18.97 fg	13.57 g
		Rain	250.1 cd	6.50 ab	19.35 f	13.03 g	18.68 g	13.29 h
	Veg	City	205.6 de	4.83 cd	19.63 de	13.86 d	19.62 de	13.88 f
		Rain	135.0 e	4.33 d	18.81 g	13.28 f	19.33 ef	13.80 f
	Mulch		<0.0001	<0.0001	<0.0001	<0.0001	<0.0001	<0.0001
	Water		0.3362	0.0576	0.0002	<0.0001	0.1511	<0.0001
	Mulch × Water		0.0718	0.4984	0.005	<0.0001	0.2181	0.0023
2014	Black	City	400.9 a	8.83 ab	18.51 a	13.17 a	19.62 a	13.98 a
		Rain	341.3 b	7.33 ab	18.26 b	12.96 b	19.34 b	13.74 b
	BlackCC	City	286.2 c	10.17 a	18.24 b	12.97 b	19.36 b	13.78 b
		Rain	293.8 c	9.33 ab	18.24 b	12.98 b	19.27 b	13.74 b
	Bare	City	281.0 c	6.00 b	17.49 c	12.25 c	18.18 c	12.89 c
		Rain	261.4 c	5.6 ab	17.11 d	11.98 d	18.13 c	12.82 c
	Mulch		<0.0001	0.1765	<0.0001	<0.0001	<0.0001	<0.0001
	Water		0.12	0.9542	0.0021	0.0072	0.0954	0.0192
	Mulch × Water		0.1956	0.2697	0.0587	0.0893	0.4201	0.217

For values within mulch, water followed by the same letter are not significantly different according to Fisher's LSD (*p* = 0.1). EF[†] and LF[†] were characterized by nighttime soil temperature in specific EF and LF.

4. Conclusions

The three-years of data demonstrated that high tunnels combined with the application of surface mulches and the row cover, along with different irrigation water can be expected to have significant impacts on soil temperature, which influenced lettuce growth and development. The mulches had significant impacts on soil temperature, which was related to total lettuce yield over the falls of 2011, 2013, and 2014. Clear and black plastics all have good soil-warming ability, thus produced the highest lettuce weights. However, the thermal stress with the higher soil temperatures produced in clear and black plastics at transplanting may have caused a reduction of plants per plot when compared to the bare. The biodegradable biobag, spunbond and paper mulches did not heat the soil as much as the clear or black mulch, but the marketable lettuce yield produced by biobag and paper mulches were not statistically different compared with the black polyethylene mulch. In addition, the cover crops incorporated into the soil underneath the black and clear mulch had the potential to save more lettuces in each plot, but overall lettuce weight per head was reduced significantly when compared to the non-cover crop treatments. The reduction in yield from the cover crop may be due to less available nitrogen for plant use, as the organic matter is being decomposed. Moreover, vegetative mulch produced the coolest soil temperatures, and consistently generated the lowest lettuce yield, so is not recommended to be applied in HTs. Additionally, during cold nights, row covers can add protection in the LF, thus produced higher lettuce production when compared to groups without row covers. Finally, the temperature of city water was generally higher than rainwater except in the EF, and it could increase the soil temperature by 0.2–0.8 °C compared to the rainwater plots. Although there were several exceptions when soil temperatures were warmer using rainwater than city water, overall lettuce production was still increased by using city water. Additionally, city water generally provided

Agriculture **2019**, *9*, 203

more nutrients than rainwater in terms of K, Mg, and Ca which may promote lettuce growth and development. Therefore, city water irrigation may not have just warmed the soil but also provided more nutrients for increased lettuce production.

Author Contributions: Conceptualization, M.Z. and B.L.; methodology, M.Z., B.L. and D.B.; software, M.Z. and B.L.; validation, M.Z. and B.L.; formal analysis, M.Z.; investigation, M.Z.; resources, B.L.; Data curation, M.Z., B.L., W.W., D.B., A.H.; writing—original draft preparation, M.Z.; writing—review and editing, M.Z., B.L., D.H., P.A., and D.H.; visualization, M.Z.; supervision, B.L.; funding acquisition, B.L.

Funding: This project was funded in part by Conservation Innovation Grants, NRCS, USDA (2013-2016) and Ag Research & Extension Innovation Grants (2011).

Conflicts of Interest: The authors declare no conflict of interest. The funders had no role in the design of the study; in the collection, analyses, or interpretation of data; in the writing of the manuscript, or in the decision to publish the results.

References

1. Hinrichs, C. The practice and politics of food system localization. *J. Rural. Stud.* **2003**, *19*, 33–45. [CrossRef]
2. Lamont, W.J. Plastics: Modifying the Microclimate for the Production of Vegetable Crops. *HortTechnology* **2005**, *15*, 477–481. [CrossRef]
3. Kittas, C.; Karamanis, M.; Katsoulas, N. Air temperature regime in a forced ventilated greenhouse with rose crop. *Energy Build.* **2005**, *37*, 807–812. [CrossRef]
4. Wells, O.S.; Loy, J.B. Rowcovers and high tunnels enhance crop production in the northeastern United States. *HortTechnology* **1993**, *3*, 92–95. [CrossRef]
5. Blomgren, T.; Frisch, T. *High Tunnels: Using Low Cost Technology to Increase Yields, Improve Quality, and Extend The Growing Season*; University of Vermont Center for Sustainable Agriculture: Burlington, VT, USA, 2007.
6. Carey, E.E.; Jett, L.; Lamont, W.J., Jr.; Nennich, T.T.; Orzolek, M.D.; Williams, K.A. Horticultural Crop Production in High Tunnels in the United States: A Snapshot. *HortTechnology* **2009**, *19*, 37–43. [CrossRef]
7. Shannon, M.; Grieve, C. Tolerance of vegetable crops to salinity. *Sci. Hortic.* **1998**, *78*, 5–38. [CrossRef]
8. Liakatas, A.; Clark, J.A.; Monteith, J.L. Measurements of the heat balance under plastic mulches: Part I. Radiation balance and soil heat flux. *Agric. For. Meteorol.* **1986**, *36*, 227–239. [CrossRef]
9. Siwek, P.; Kalisz, A.; Wojciechowska, R. Effect of mulching with film of different colours made from original and recycled polyethylene on the yield of butterhead lettuce and celery. *Folia Hort. Supl.* **2007**, *19*, 25–35.
10. Iqbal, Q.; Amjad, M.; Asi, M.R.; Ali, M.A.; Ahmad, R. Vegetative and reproductive evaluation of hot peppers under different plastic mulches in poly/plastic tunnel. *Pak. J. Agri. Sci.* **2009**, *46*, 113–118.
11. Kapanen, A.; Schettini, E.; Vox, G.; Itävaara, M. Performance and Environmental Impact of Biodegradable Films in Agriculture: A Field Study on Protected Cultivation. *J. Polym. Environ.* **2008**, *16*, 109–122. [CrossRef]
12. Kasirajan, S.; Ngouajio, M. Polyethylene and biodegradable mulches for agricultural applications: A review. *Agron. Sustain. Dev.* **2012**, *32*, 501–529. [CrossRef]
13. Olsen, J.K.; Gounder, R.K. Alternatives to polyethylene mulch film—A field assessment of transported materials in capsicum (*Capsicum annuum* L.). *Aust. J. Exp. Agric.* **2001**, *41*, 93. [CrossRef]
14. Moreno, M.; Moreno, A. Effect of different biodegradable and polyethylene mulches on soil properties and production in a tomato crop. *Sci. Hortic.* **2008**, *116*, 256–263. [CrossRef]
15. Miles, C.; Wallace, R.; Wszelaki, A.; Martin, J.; Cowan, J.; Walters, T.; Inglis, D. Deterioration of Potentially Biodegradable Alternatives to Black Plastic Mulch in Three Tomato Production Regions. *HortScience* **2012**, *47*, 1270–1277. [CrossRef]
16. Brault, D.; Stewart, K.; Jenni, S. Optical Properties of Paper and Polyethylene Mulches Used for Weed Control in Lettuce. *HortScience* **2002**, *37*, 87–91. [CrossRef]
17. Ibarra-Jiménez, L.; Quezada-Martín, M.R.; De La Rosa-Ibarra, M. The effect of plastic mulch and row covers on the growth and physiology of cucumber. *Aust. J. Exp. Agric.* **2004**, *44*, 91–94. [CrossRef]
18. Gimenez, C.; Otto, R.; Castilla, N. Productivity of leaf and root vegetable crops under direct cover. *Sci. Hortic.* **2002**, *94*, 1–11. [CrossRef]
19. Teasdale, J.R.; Daughtry, C.S.T. Weed Suppression by Live and Desiccated Hairy Vetch (*Vicia villosa*). *Weed Sci.* **1993**, *41*, 207–212. [CrossRef]

20. Liebman, M.; Davis, A.S. Integration of soil, crop and weed management in low-external-input farming systems. *Weed Res.* **2000**, *40*, 27–47. [CrossRef]

21. Zheng, M.; Leib, B.; Butler, D.; Wright, W.; Ayers, P.; Hayes, D.; Haghverdi, A.; Feng, L.; Grant, T.; Vanchiasong, P.; et al. Assessing heat management practices in high tunnels to improve organic production of bell peppers. *Sci. Hortic.* **2019**, *246*, 928–941. [CrossRef]

agriculture

MDPI

Brief Report

Adjustment of Irrigation Schedules as a Strategy to Mitigate Climate Change Impacts on Agriculture in Cyprus

Panagiotis Dalias *, Anastasis Christou and Damianos Neocleous

Agricultural Research Institute, Ministry of Agriculture, Rural Development and Environment, P.O. Box 22016, 1516 Nicosia, Cyprus; Anastasis.Christou@ari.gov.cy (A.C.); d.neocleous@ari.gov.cy (D.N.)
* Correspondence: dalias@ari.gov.cy; Tel.: +357-22403114; Fax: +357-22316770

Received: 30 November 2018; Accepted: 18 December 2018; Published: 21 December 2018

Abstract: The study aimed at investigating eventual deviations from typical recommendations of irrigation water application to crops in Cyprus given the undeniable changes in recent weather conditions. It focused on the seasonal or monthly changes in crop evapotranspiration (ETc) and net irrigation requirements (NIR) of a number of permanent and annual crops over two consecutive overlapping periods (1976–2000 and 1990–2014). While the differences in the seasonal ETc and NIR estimates were not statistically significant between the studied periods, differences were identified via a month-by-month comparison. In March, the water demands of crops appeared to be significantly greater during the recent past in relation to 1976–2000, while for NIR, March showed statistically significant increases and September showed significant decreases. Consequently, the adjustment of irrigation schedules to climate change by farmers should not rely on annual trends as an eventual mismatch of monthly crop water needs with irrigation water supply might affect the critical growth stages of crops with a disproportionately greater negative impact on yields and quality. The clear increase in irrigation needs in March coincides with the most sensitive growth stage of irrigated potato crops in Cyprus. Therefore, the results may serve as a useful tool for current and future adaptation measures.

Keywords: climate change adaptation; irrigated crops; net irrigation requirements; crop evapotranspiration; monthly changes

---

1. Introduction

Theoretical considerations, climate simulation models and empirical evidence indicate that global warming is leading to increased water vapor and to increased land precipitation at higher latitudes, notably over North America and Eurasia [1]. However, contrary to many mid-to-upper latitude regions of the world, several regional studies have shown a dominant decreasing trend over the Mediterranean Basin [2], although changes will not be equivalent across all Mediterranean regions or seasons [3].

In Cyprus, this change is already being manifested by a decrease in mean annual rainfall and an increase in annual mean temperature. Model projections agree on its future warming and drying, with a likely increase of heatwaves and dry spells; a prospect that will worsen the already existing water scarcity [4,5].

The consequences of such temperature and precipitation changes on a number of aspects of human life and agriculture might be considerable. In agriculture, increased temperatures or the extension of dryness may have a negative impact on crop yields [6] and in turn on food security [7] and may influence crops dynamics, e.g., the exclusion of some crops, or their replacement by others more adapted to the new conditions [8]. Changes in climatic conditions might also affect the proliferation and spread of invasive species, weeds, or diseases [9].

Crop production in Cyprus is covered by annual (e.g., potatoes and vegetables) and permanent (e.g., citrus, olives and grapevines) crops summing over 100×10^3 ha, of which 30% is irrigable land. Citrus and potatoes are the most widely grown crops in the country and consume over 30% of the total agricultural water (150×10^6 m^3). Crop production is constrained by a highly variable climate, limited precipitation and high temperatures from mid-May to mid-September [10]. However, crop water needs may be fully or partly met by rainfall mainly from October to March. Given the projected lower precipitation it can reasonably be assumed that irrigation water availability and crop yields will be affected [11]. Nonetheless, previous work showed that, considering the changes over recent years in mean rainfall and pan evaporation data, the total irrigation needs of crops in Cyprus have not been modified, at least until now [12].

While extreme weather events, which are predicted to increase under future climate scenarios, are already considered a significant challenge for producers [13], little work has been done so far on the current seasonal or monthly changes of temperature or the distribution of precipitation throughout the year and the consequences that these modifications may bring upon crops. Such changes may have an impact on some critical stages of the biological cycle of plants and disproportionately affect productivity and yields [14]. For example, irrigation experiments showed different effects on wheat yield, quality, and water-use efficiencies depending on the plant-growth or phenological stage at which water deficits were applied [15,16]. Therefore, an eventual mismatch of crop water needs with irrigation water supply might be critical, and adaptation measures related to irrigation schedules or the adjustment of planting/seeding dates might be necessary.

This study aims to investigate whether one of the characteristics of the ongoing climate change in Cyprus is a significant modification of the seasonal or monthly water needs and irrigation demands of crops, and discusses the consequences for agricultural production of an eventual deviation of the prevailing irrigation schedules to the current climatic conditions. It also investigates the possibilities for adaptation to climate change challenges using planting period shifts or irrigation schedule modifications.

2. Materials and Methods

The analysis of water and irrigation needs in this study was applied to 35 irrigated crops cultivated in Cyprus. Some of these crops require their water needs to be met fully or partially by irrigation, while some require irrigation only occasionally.

Mean monthly precipitation (mm) and pan evaporation (screened USWB Class A pan) data from 16 weather stations provided by the Cyprus Meteorological Department were used. These stations were situated in the main agricultural areas of Cyprus.

Crop evapotranspiration (ETc) was calculated from potential evapotranspiration (ETo) and pan evaporation (Epan) data obtained from the weather stations using the methodology proposed by the Food and Agriculture Organization of the United Nations (FAO) [17]. More precisely, Epan measurements were converted to reference evapotranspiration (ETo) using the equation:

$$ETo = Kp \times Epan \tag{1}$$

where Kp is the pan coefficient, which takes into account the type of pan, its environment and climate. Potential crop evapotranspiration (ETc) was calculated from ETo according to:

$$ETc = Kc \times ETo \tag{2}$$

where Kc is the crop coefficient, which depends on the kind of crop and its stage of development. Combining the previous two equations, ETc can be expressed as:

$$ETc = Kc \times Kp \times Epan \tag{3}$$

Substituting Kc × Kp with a coefficient C equation, (3) becomes:

$$ETc = C \times Epan \tag{4}$$

i.e., crop evapotranspiration is calculated directly from pan evaporation using a single coefficient. Values of the coefficients can be derived from the literature [18,19] but they were extensively studied and adjusted to local conditions by the Agricultural Research Institute of Cyprus (e.g., [20,21]).

The net irrigation requirements (NIR) of the crops were calculated by subtracting from the actual water requirements of crops (ETc values) the effective rainfall (Pe), i.e., rain water that is not percolated below the root zone or run-off, but is stored in the root zone and can be used by the plants [17].

ETc and NIR values were estimated for the 35 crops cultivated in Cyprus under two overlapping periods, 1976–2000 and 1990–2014, for each of the 16 weather stations. The mean values obtained for these two periods for each season (winter: December–January–February, spring: March–April–May, summer: June–July–August, and autumn: September–October–November) and each month were compared for each crop.

Statistical Analysis

The mean seasonal or monthly ETc and NIR values for each studied crop and mean seasonal or monthly precipitation and Epan values obtained from the 16 weather stations were compared between the two studied periods (1976–2000 and 1990–2014) using a paired *t*-test (16 double samples for each crop over the 16 weather stations). A *p*-value < 0.05 in these tests was considered as statistically significant (GraphPad Software, Inc., San Diego, CA, USA).

3. Results

Differences of seasonal ETc and NIR estimates were not statistically significant between the studied periods. The results of the comparison of seasonal NIR between the two studied periods for selected crops cultivated in Cyprus are shown in Table 1. All crops showed decreased average irrigation needs in all seasons apart from spring, where some crops appeared to have greater irrigation water demands in the recent past. The *p* values for the NIR of many crops were close to statistical significance in autumn.

Table 1. *p*-Values for the comparison of seasonal net irrigation requirements (NIR) between the two periods (1976–2000 and 1990–2014) for selected crops cultivated in Cyprus. Months that are not included in the irrigation period of a crop are indicated by n/a (non-applicable).

	Winter	Spring	Summer	Autumn
Fruit trees (mountains)	n/a	0.7157	0.1989	0.0680
Green beans: greenhouse	0.1132	0.6404	n/a	0.1973
Haricot beans	n/a	n/a	n/a	0.0728
Lettuce	0.6292	n/a	n/a	0.1639
Marrows: outside grown	n/a	0.6714	0.1924	n/a
Melons: outside grown	n/a	0.6714	0.1879	n/a
Monkey nuts	n/a	0.6080	0.1667	0.0680
Okra (lady's fingers)	n/a	0.9315	0.1750	n/a
Onions dried	n/a	0.9814	0.5567	n/a
Peas general	n/a	0.8110	n/a	n/a
Peppers: outside grown	n/a	0.6533	0.1773	0.0680
Pistachio	n/a	n/a	0.1903	0.0680
Potatoes (spring crop)	n/a	0.8070	n/a	n/a
Radish	n/a	0.0054	n/a	0.0959
Spinach	n/a	n/a	n/a	0.0959
Table grapes	n/a	0.6484	0.5567	n/a
Table olives	n/a	0.6271	0.1953	0.0712

Table 2, showing the results for only the most water-consuming crops, indicates that March was the only month in which there was a statistically significant difference (increase) in ETc between the two periods ($p < 0.05$). The water demand of all crops in this month was significantly greater in recent years than in the distant past. Void cells in Table 2 indicate months that are not included in the irrigation period of crops in Cyprus. For NIR, apart from March, statistically significant differences between the two periods were also found for September, as seen in Table 3. In March, the irrigation requirements were greater for 1990–2014 than for 1976–2000, in contrast to what was found for September.

Table 2. Crop evapotranspiration (ETc) values in mm of the most water-consuming crops in Cyprus. The upper number for each month indicates the average value for the 1976–2000 period and the lower number indicates the average value for the period 1990–2014. Non-significant differences between these two averages are indicated by n.s. (paired *t*-test) and significant differences ($p < 0.05$) by *. Months that are not included in the irrigation period of a crop are indicated by n/a (non-applicable).

	Bananas (*Musa* spp.)		Citrus (*Citrus* spp.)		Taro (*Colocasia esculenta*)		Potatoes (*Solanum tuberosum*)	
January	n/a		n/a		n/a		n/a	
February	n/a		n/a		n/a		n/a	
March	22.9	*	18.3	*	32.9	*	54.9	*
	23.9		19.1		34.4		57.4	
April	70.0	n.s.	65.2	n.s.	157.3	n.s.	95.9	n.s.
	69.3		64.6		155.7		94.9	
May	116.4	n.s.	99.7	n.s.	186.3	n.s.	130.4	n.s.
	115.9		99.2		185.5		129.8	
June	174.1	n.s.	132.3	n.s.	378.0	n.s.	n/a	
	173.1		131.6		375.9			
July	219.4	n.s.	138.3	n.s.	448.3	n.s.	n/a	
	214.6		135.3		438.5			
August	231.8	n.s.	176.0	n.s.	452.0	n.s.	n/a	
	227.2		172.5		443.1			
September	195.5	n.s.	119.4	n.s.	365.9	n.s.	n/a	
	189.0		115.5		353.8			
October	123.5	n.s.	52.7	n.s.	153.2	n.s.	n/a	
	119.7		51.0		148.5			
November	49.2	n.s.	9.6	n.s.	135.0	n.s.	n/a	
	47.9		9.4		131.5			
December	n/a		n/a		n/a		n/a	

Meteorological precipitation and evaporation data was analyzed to gain insight into the causes of this change. Climate charts of the distribution of rainfall and Epan over the months of the year (Figure 1a,b) illustrate the differences in these meteorological variables between the two 24-year periods. The mean March Epan value for 1990–2014 was significantly increased in relation to the 1976–2000 interval. March and September were the only months that this statistically significant difference was observed. An increase in Epan in March was recorded at 13 of 16 weather stations.

All 16 stations showed decreased average precipitation in more recent years, with an average reduction of 36%. In all other months, the stations showed both increases and decreases in average rainfall when the two periods were compared. The statistically strong tendency to decreased precipitation during March was not followed by a respective decrease during the following two months of spring. In May, for example, at 12 out of 16 stations a tendency towards an increase in precipitation was noted. For September, which also showed a statistically significant change in the irrigation needs of crops, the opposite trend was manifested, with only 1 station out of 16 recording a decrease in rainfall.

a)

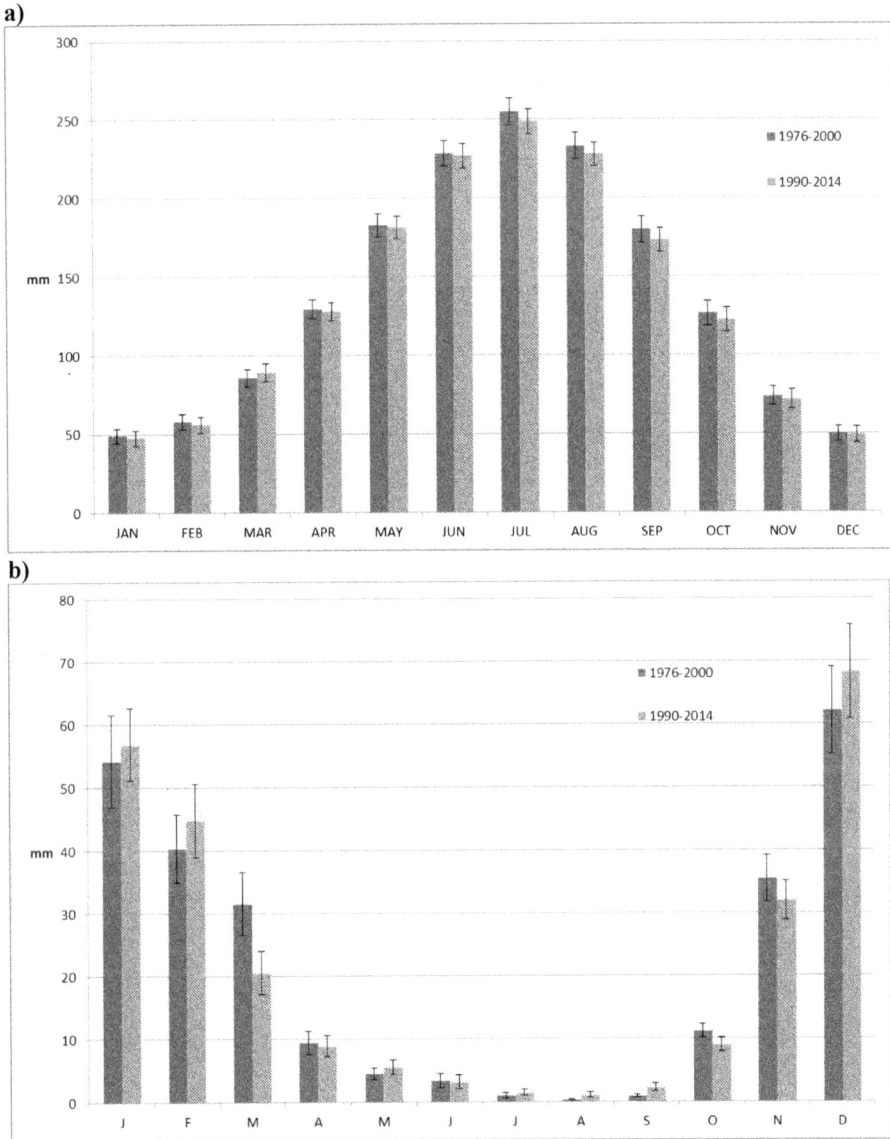

b)

Figure 1. Monthly averages for (**a**) Class A pan evaporation (mm) and (**b**) effective rainfall over two 24-year periods (1976–2000 and 1990–2014) in Cyprus. Data from 16 meteorological weather stations.

Table 3. Net irrigation requirements (NIR) values in mm of the most water-consuming crops in Cyprus. The upper number for each month indicates the average value for the 1976–2000 period and the lower number the average value for the period 1990–2014. Non-significant differences between these two averages are indicated by n.s. (paired *t*-test) and significant differences (*p* < 0.05) by *. Months that are not included in the irrigation period of a crop are indicated by n/a (non-applicable).

	Bananas (*Musa* spp.)		Citrus (*Citrus* spp.)		Taro (*Colocasia esculenta*)		Potatoes (*Solanum tuberosum*)	
January	n/a		n/a		n/a		n/a	
February	n/a		n/a		n/a		n/a	
March	8.1 13.2	*	5.0 8.1	*	15.4 21.3	*	34.5 39.9	*
April	64.7 64.5	n.s.	59.5 59.4	n.s.	157.7 156.6	n.s.	92.3 91.8	n.s.
May	119.5 117.9	n.s.	101.6 100.0	n.s.	194.0 192.0	n.s.	134.4 132.7	n.s.
June	182.1 181.2	n.s.	137.5 136.9	n.s.	399.6 397.6	n.s.	n/a	
July	218.4 213.0	n.s.	137.3 133.7	n.s.	447.3 436.9	n.s.	n/a	
August	231.5 226.1	n.s.	175.7 171.4	n.s.	451.7 442.0	n.s.	n/a	
September	194.6 186.7	*	118.5 113.2	*	365.0 351.5	*	n/a	
October	120.0 118.2	n.s.	44.4 45.0	n.s.	151.7 148.9	n.s.	n/a	
November	20.1 20.3	n.s.	n/a	n.s.	107.9 107.5	n.s.	n/a	
December	n/a		n/a		n/a		n/a	

4. Discussion

The re-estimation of irrigation required two successive past periods in order to evaluate the effect of the ongoing changes in precipitation and evaporative demand of the atmosphere on the water demand of crops. The results revealed some interesting effects of ongoing climate change, which usually do not receive the deserved attention, and which could prove to be a useful guide for farmers, policy makers, government officers and agricultural advisors.

The trends in the change of mean annual precipitation and mean annual temperature in Cyprus are not reflected equally or proportionally at different times of the year. Consequently, the adjustment of irrigation schedules to climate change by farmers should not rely on annual trends as practiced by local growers. Focusing on month-by-month changes revealed strong trends towards an increase in evaporation during March at all meteorological stations, which in combination with a respective decrease in precipitation attests that an adjustment of irrigation water provision to crops is needed. Irrigation programs that are based on "old" meteorological data would result in water deficiencies, which may affect critical growth stages of plants. Moreover, in many cases, farmers now need to irrigate their crops during March, whereas previously, irrigation in March was negligible. In March, the precipitation dropped by 36% and the amount of water that would be needed to compensate for this reduction was estimated to be also 36% on average, as rainfall covered a large part of the total water demand of crops. Climate change effects on irrigation scheduling parameters were also found in Calabria, Italy. From an analysis of reference evapotranspiration (ETo) during the last decades, it was shown that a positive trend in summer precipitation also caused an advance of the last watering, resulting in a slight decrease of the length of the irrigation season [22].

The example of potatoes is probably indicative of the necessary adaptation measures. Potatoes are one of the most exportable products of Cyprus and one of the most water-consuming crops. The "spring crop" or the "main crop" is planted in November/February and harvested in March/June mainly for export, but also for local consumption. Stolonization and tuber initiation are the stages that are most sensitive to water shortage [23], mainly because they are the stages of the highest crop water demand. If water shortages occur during the mid-season stage, which in Cyprus coincides with March, the negative effect on the yield will be pronounced. Karafyllidis et al. [24] showed that limited soil moisture availability affected yield and the number and size of tubers. In the following year, seed produced under conditions of moisture stress produced plants with 20% fewer stems, 24–33% less yield, 18–22% fewer tubers and 19–22% fewer large tubers than plants from seed produced under abundant water supply.

Hence, irrigation should be applied as an adaptation measure to safeguard yields if meteorological trends continue as they are today. An earlier shift of plantation dates could alternatively also be envisaged as an adaptation measure of potato cultivation, as crops would have completed their water-sensitive stage before the less favorable conditions of March. An analysis of the optimum adjustment of planting dates for corn and soybean was also suggested by Woznicki et al. [25] as one of the best adaptation strategies to cope with future climate change scenarios.

However, in contrast, precipitation increased from 0.9 to 2.3 mm in September, affecting irrigation water demand for this month by only 5%. This is because the contribution of rainfall to the total amount of water that needed to be applied to crops in September was nevertheless very small. In this case, following current irrigation guidelines would result in supplying crops with an excess of water. This would not have a negative effect on productivity and yields but it would result in wasting water. The results, therefore, support the notion that in changing climatic conditions, the irrigation adaptation actions required are different in each case depending on specific conditions. Using a modeling approach to simulate the impact of various climate change scenarios on crop water and downscaling climatic parameters derived from global circulation models, Doria and Madramootoo [26] similarly suggested that in order to sustain crop production in the future, efficient irrigation scheduling for producers should be used as an adaptation measure.

The monthly changes in weather conditions that were highlighted in this study and their significant effects on agricultural production constitute a very subtle aspect of climate change, as they are not obvious even as seasonal changes. As a result, we advocate for further examination and verification in other places with a similar climate. However, if the shown precipitation and evaporation trends continue in the future, rainfed crops could also be affected and emphasis should be placed on supplementary irrigation during March. The addition of small amounts of water in this month could improve and stabilize yields, providing the missing moisture for normal plant growth.

5. Conclusions

Irrigation schedules that are based on the average evaporation and rainfall records of an area have to be adjusted to recent changes of climatic parameters even if the year-round changes are not significantly affected. Shifts in rainfall and temperature "allocation" across the months of the year call for a corresponding adjustment of the irrigation water applied to crops, as an eventual mismatch with plant needs could significantly affect some of their critical growth stages. The adjustment of irrigation schedules should be based on more local studies, even if they are in opposition to trends found in wider areas.

Author Contributions: Each author made substantial contributions to this publication. P.D. collected and analyzed the data and wrote the first draft. D.N. and A.C. had a significant contribution to the improvement and revision of the manuscript.

Funding: This research received no external funding.

Acknowledgments: This work was supported by the Agricultural Research Institute of Cyprus (ARI) and authors did not receive any specific grant from funding agencies in the public, commercial, or not-for-profit sectors. The authors would like to thank the staff of the Natural Resources and Environment Section of ARI for their assistance.

Conflicts of Interest: The authors declare no conflict of interest.

References

1. Trenberth, K.E.; Jones, P.D.; Ambenje, P.; Bojariu, R.; Easterling, D.; Klein Tank, A.; Parker, D.; Rahimzadeh, F.; Renwick, J.A.; Rusticucci, M.; et al. Observations: Surface and atmospheric climate change. In *Climate Change 2007: The Physical Science Basis*; Solomon, S., Qin, D., Manning, M., Chen, Z., Marquis, M., Averyt, K.B., Tignor, M., Miller, H.L., Eds.; Intergovernmental Panel on Climate Change 4th Assessment Report; Cambridge University Press: Cambridge, UK, 2007; pp. 235–336.
2. Alpert, P.; Ben-Gai, T.; Baharad, A.; Benjamini, Y.; Yekutieli, D.; Colacino, M.; Diodato, L.; Ramis, C.; Homar, V.; Romero, R.; et al. The paradoxical increase of Mediterranean extreme daily rainfall in spite of decrease in total values. *Geophys. Res. Lett.* **2002**, *29*, 1536. [CrossRef]
3. Misra, A.K. Climate change and challenges of water and food security. *Int. J. Sustain. Built Environ.* **2014**, *3*, 153–165. [CrossRef]
4. Lelieveld, J.; Hadjinicolaou, P.; Kostopoulou, E.; Chenoweth, J.; El Maayar, M.; Giannakopoulos, C.; Hannides, C.; Lange, M.A.; Tanarhte, M.; Tyrlis, E.; et al. Climate change and impacts in the eastern Mediterranean and the Middle East. *Clim. Chang.* **2012**, *114*, 667–687. [CrossRef] [PubMed]
5. Lionello, P.; Abrantes, F.; Gacic, M.; Planton, S.; Trigo, R.; Lbrich, U. The climate of the Mediterranean region: Research progress and climate change impacts. *Reg. Environ. Chang.* **2014**, *14*, 1679–1684. [CrossRef]
6. Lobell, D.B.; Field, C.B. Global scale climate–crop yield relationships and the impacts of recent warming. *Environ. Res. Lett.* **2007**, *2*, 014002. [CrossRef]
7. Schmidhuber, J.; Tubiello, F.N. Global food security under climate change. *Proc. Natl. Acad. Sci. USA* **2007**, *104*, 19703–19708. [CrossRef]
8. Olesen, J.E.; Bindi, M. Consequences of climate change for European agricultural productivity, land use and policy. *Eur. J. Agron.* **2002**, *16*, 239–262. [CrossRef]
9. Rosenzweig, C.; Iglesias, A.; Yang, X.B.; Epstein, P.R.; Chivian, E. Climate change and extreme weather events—Implications for food production, plant diseases, and pests. *Glob. Chang. Hum. Health* **2001**, *2*, 90–104. [CrossRef]
10. Papadavid, G.; Neocleous, D.; Kountios, G.; Markou, M.; Michailidis, A.; Ragkos, A.; Hadjimitsis, D. Using SEBAL to Investigate How Variations in Climate Impact on Crop Evapotranspiration. *J. Imaging* **2017**, *3*, 30. [CrossRef]
11. Fraga, H.; Carcia de Cortázar Atauri, I.; Santos, J.A. Viticulture irrigation demands under climate change scenarios in Portugal. *Agric. Water Manag.* **2018**, *196*, 66–74. [CrossRef]
12. Christou, A.; Dalias, P.; Neocleous, D. Spatial and temporal variations in evapotranspiration and net water requirements of typical Mediterranean crops on the island of Cyprus. *J. Agric. Sci.* **2017**, *155*, 1311–1323. [CrossRef]
13. Barlow, K.M.; Christy, B.P.; OLeary, G.J.; Riffkin, P.A.; Nuttall, J.G. Simulating the impact of extreme heat and frost events on wheat crop production: A review. *Field Crops Res.* **2015**, *171*, 109–119. [CrossRef]
14. Hatfield, J.L.; Prueger, J.H. Temperature extremes: Effect on plant growth and development. *Weather Clim. Extremes* **2015**, *10*, 4–10. [CrossRef]
15. Ali, M.H.; Hoque, M.R.; Hassan, A.A.; Khair, A. Effects of deficit irrigation on yield, water productivity, and economic returns of wheat. *Agric. Water Manag.* **2007**, *92*, 151–161. [CrossRef]
16. Tari, A.F. The effects of different deficit irrigation strategies on yield, quality, and water-use efficiencies of wheat under semi-arid conditions. *Agric. Water Manag.* **2016**, *167*, 1–10. [CrossRef]
17. Allen, R.G.; Pereira, L.S.; Raes, D.; Smith, M. *Crop Evapotranspiration—Guidelines for Computing Crop Water Requirements*; FAO Irrigation and Drainage Paper No. 56; FAO: Rome, Italy, 1998.
18. Doorenbos, J.; Pruitt, W.O. *Guidelines for Predicting Crop Water Requirements*; FAO Irrigation and Drainage Paper No. 24; FAO: Rome, Italy, 1975.
19. Doorenbos, J.; Kassam, A.H. *Yield Response to Water*; FAO Irrigation and Drainage Paper No. 33; FAO: Rome, Italy, 1979.
20. Metochis, C. Irrigation of lucerne under semi-arid conditions in Cyprus. *Irrig. Sci.* **1980**, *1*, 247–252. [CrossRef]
21. Metochis, C. Water requirement, yield and fruit quality of grapefruit irrigated with high-sulphate water. *J. Hortic. Sci.* **1989**, *64*, 733–737. [CrossRef]

22. Capra, A.; Mannino, R. Effects of climate change on the irrigation scheduling parameters in Calabria (south Italy) during 1925–2013. *Irrig. Drain. Syst. Eng.* **2015**, *S1*, 003.

23. Brouwer, C.; Prins, K.; Heibloem, M. *Irrigation Water Management: Irrigation Scheduling*; Training Manual No. 4; FAO: Rome, Italy, 1989.

24. Karafyllidis, D.I.; Stavropoulos, N.; Georgakis, D. The effect of water stress on the yielding capacity of potato crops and subsequent performance of seed tubers. *Potato Res.* **1996**, *39*, 153–163. [CrossRef]

25. Woznicki, S.A.; Pouyan Nejadhashemi, A.; Parsinejad, M. Climate change and irrigation demand: Uncertainty and adaptation. *J. Hydrol. Reg. Stud.* **2015**, *3*, 247–264. [CrossRef]

26. Doria, R.O.; Madramootoo, C.A. Estimation of irrigation requirements for some crops in southern Quebec using CROPWAT. *Irrig. Drain.* **2009**, *64*, 1–11.

MDPI

St. Alban-Anlage 66

4052 Basel

Switzerland

Tel. +41 61 683 77 34

Fax +41 61 302 89 18

www.mdpi.com

Agriculture Editorial Office

E-mail: agriculture@mdpi.com

www.mdpi.com/journal/agriculture

www.ingramcontent.com/pod-product-compliance
Lightning Source LLC
Chambersburg PA
CBHW041215220326
41597CB00033BA/5975